Building a Platform for Data-Driven Pandemic Prediction

Building a Platform for Data-Driven Pandemic Prediction

From Data Modelling to Visualisation - The CovidLP Project

Edited by
Dani Gamerman
Marcos O. Prates
Thaís Paiva
Vinícius D. Mayrink

CRC Press
Taylor & Francis Group
Boca Raton London New York

CRC Press is an imprint of the
Taylor & Francis Group, an **informa** business

A CHAPMAN & HALL BOOK

First edition published 2022
by CRC Press
6000 Broken Sound Parkway NW, Suite 300, Boca Raton, FL 33487-2742

and by CRC Press
2 Park Square, Milton Park, Abingdon, Oxon, OX14 4RN

© 2022 selection and editorial matter, Dani Gamerman, Marcos O. Prates, Thaís Paiva, Vinícius D. Mayrink; individual chapters, the contributors

CRC Press is an imprint of Taylor & Francis Group, LLC

Library of Congress Cataloging-in-Publication Data

ISBN: 978-0-367-70999-0 (hbk)
ISBN: 978-0-367-70997-6 (pbk)
ISBN: 978-1-003-14888-3 (ebk)

DOI: 10.1201/9781003148883

Typeset in [font]
by KnowledgeWorks Global Ltd.

To Science

Contents

Preface

The project that led to this book started in March 2020. One of us was just starting his graduate course classes on Dynamic models when the COVID-19 pandemic started in Brazil, forcing the suspension of presential classes. During the first months of the pandemic, universities in Brazil were unsure of how to proceed. Our university recommended that faculty should not continue classes even in on-line mode and should resort only to challenges to the students and basic exercises.

About that time, an intense virtual debate started among groups of faculty. This pattern was observed in our Institute of Exact Sciences, which consists of the departments of Statistics, Mathematics, Computing, Physics and Chemistry. New messages containing solutions to various problems relating to the pandemic appeared every single day from members of all the departments.

One of the messages contained a data-driven proposal based on the logistic curve, with parameter estimation and prediction of the counts of new cases until the pandemic ends. The message was written by a physicist and contained an abridged version of a paper. This manuscript was the spark that was missing to decide on how to entertain the students. After all, statisticians should be able to handle at least the task presented by the Physics colleague.

The project was presented to the graduate students, who were very keen on embracing the exercise and started working straight away on the project. Their results started to appear and problems started to emerge. They were discussed at regular meetings held at class time. After all, this was a challenge for the student and this was allowed by the university rules to replace the missing classes!

Our results were regularly passed on informally to our departmental colleagues. They always pointed to the need to inform the general public about what the project was providing. This issue led to the need to build an appropriate platform for the release of the project information. One of us was drawn to the project at this point. The preparation of the platform for releasing the results led to the CovidLP app.

Round about the same time, the Ministry of Education opened an urgent call for proposals on different research aspects of the pandemic. Our project was submitted and subsequently approved: 2 post-doc grants and 2 workstations were dedicated to the project, giving the project the respective methodological and the computational amplitudes it so badly needed. Two other faculty and a former graduate student were also included. Our project gained the scalability it required. The CovidLP project was created. The name was chosen to emphasise the interest of the project on long-term prediction.

A number of issues of all sorts appeared and were dealt with in the best possible manner. These issues ranged from installing the workstations manually and remotely in a deserted campus to addressing the methodological difficulties and testing the proposed solutions. Many hours were spent studying the literature and testing different approaches.

The CovidLP project gained national visibility after a series of workshop and seminar presentations, media interviews and news releases, and our methodology was adopted by a few health officials in different administrative levels in Brazil. By then, the project also contained a site and a blog to inform the general public about the changes that were being introduced and discuss them. It became obvious that another stream worth pursuing was software for reproducing the analyses in a more general setting, suited for more experienced data analysts. It also became clear that the project was not restricted to the COVID-19 pandemic. Thus the software was named PandemicLP, to signal this change in scope.

After a few months, the project was mature. The participants were organised in focal groups and the production of information became more effective. The project was getting ready for scientific publication. It became clear to us that what we had developed thus far was worth reporting.

But it was clear to us that our story was not of Statistics. There are better books already available to describe statistical aspects of epidemics. It was also clear that our story was not of Computing. Again, there are better books about the capabilities now widely available in software. The story of the CovidLP project is a tale about the inseparable roles played by Statistics and Computing for building such platforms for daily release of the results of statistical analyses. This is the story that we want to tell.

By that time, CRC released a call to the scientific community for proposals in general, including books, about the pandemic. It seemed to us the perfect match, and after revising the proposal to include very thoughtful comments from reviewers, whom we thank, the proposal was accepted. A major contribution from this revision process was the recommendation for inclusion of more epidemiological background on data and models. This was accepted and this addition was provided by colleagues outside (but aware of) our project, providing an important complement to our work. Our final proposal contains the key elements of our task: data description, statistical modelling and monitoring, computational implementation and software.

We worked very hard during the 9 months that elapsed between then and now and are very pleased with our end result. Of course, this is way too short. The main intention of the book is to provide data analysts with the tools required for building platforms for statistical analyses and predictions in epidemiological contexts. A secondary goal is to allow users to make adaptations in the book structure to guide them into an online platform solution to their own data analysis problem, which may not even be related to Epidemiology.

We would like to finish by thanking the people who accompanied the development of the CovidLP project. These include the users of our app, the attendees at the talks we delivered, our academic colleagues who provided useful inputs to the project and friends and families that provided us support for achievement of this task. A very special thank you goes to the CovidLP team, a group of dedicated students and post-docs that embarked on the journey that led to this book, reading countless papers and books, implementing a number of computational codes and participating actively in the all steps required for the completion of this book. Thanks are also due to Ricardo Pedroso, for the book cover figure. A warm acknowledgement goes to Rob Calver for his continued support and relentless effort to make this book possible in the best possible shape. We also thank the CRC team, especially Rob, Michele and Vaishali, for the administrative support. They also arranged for one text editor and 2 experts to review the entire book. Their reviews, which we gratefully thank, provided more context and breadth to the book content.

In Chapter 1, the book provides different paths to follow in order to help its readers achieving their own, different goals. If the book manages to help readers to attain their platform building goals, then the book would have achieved its goals.

DG, MOP, TP and VDM

Belo Horizonte, 31 May 2021.

Contributors

The CovidLP Team
Universidade Federal de Minas Gerais
Belo Horizonte, Brazil

The team has the following members: Gabriel O. Assunção, Douglas R. M. Azevedo, Ana Julia A. Câmara, Marta Cristina C. Bianchi, Juliana Freitas, Dani Gamerman, Débora F. Magalhães, Jonathan S. Matias, Vinícius D. Mayrink, Thais P. Menezes, Guido A. Moreira, Leonardo Nascimento, Thaís Paiva, Ricardo C. Pedroso and Marcos O. Prates

Gabriel O. Assunção
Universidade Federal de Minas
  Gerais
Belo Horizonte, Brazil

Douglas R. M. Azevedo
Localiza S.A.
Belo Horizonte, Brazil

Leonardo S. Bastos
Fundação Oswaldo Cruz
Rio de Janeiro, Brazil

Ana Julia A. Câmara
Universidade Federal de Minas
  Gerais
Belo Horizonte, Brazil

Luiz M. Carvalho
Fundação Getulio Vargas
Rio de Janeiro, Brazil

Marta Cristina C. Bianchi
Universidade Federal de Minas
  Gerais
Belo Horizonte, Brazil

Juliana Freitas
Universidade Federal de Minas
  Gerais
Belo Horizonte, Brazil

Dani Gamerman
Universidade Federal de Minas
  Gerais/Universidade Federal do
  Rio de Janeiro
Belo Horizonte/Rio de Janeiro,
  Brazil

Marcelo F. C. Gomes
Fundação Oswaldo Cruz
Rio de Janeiro, Brazil

Débora F. Magalhães
Universidade Federal de Minas
  Gerais
Belo Horizonte, Brazil

Jonathan S. Matias
Universidade Federal de Minas
  Gerais
Belo Horizonte, Brazil

Vinícius D. Mayrink
Universidade Federal de Minas
  Gerais
Belo Horizonte, Brazil

Thais P. Menezes
University College
Dublin, Ireland

Guido A. Moreira
Universidade Federal de Minas
  Gerais
Belo Horizonte, Brazil

Leonardo Nascimento
Universidade Federal do Amazonas
Manaus, Brazil

Thaís Paiva
Universidade Federal de Minas
  Gerais
Belo Horizonte, Brazil

Ricardo C. Pedroso
Universidade Federal de Minas
  Gerais
Belo Horizonte, Brazil

Marcos O. Prates
Universidade Federal de Minas
  Gerais
Belo Horizonte, Brazil

Part I

Introduction

1

Overview of the book

Dani Gamerman

Universidade Federal de Minas Gerais/Universidade Federal do Rio de Janeiro, Brazil

Thaís Paiva

Universidade Federal de Minas Gerais, Brazil

Guido A. Moreira

Universidade Federal de Minas Gerais, Brazil

Juliana Freitas

Universidade Federal de Minas Gerais, Brazil

CONTENTS

This is a book about building platforms for pandemic prediction. In order to predict pandemics and epidemics, it is necessary to develop an inferential system typically based on Statistics. In order to build platforms, it is necessary to develop tools typically based on Computing. Both parts are important in the development of a platform and will be treated with equal importance. The book is structured in parts that will handle each component of this construction process, and describes their integration. This structure aims to benefit readers interested in building platforms and/or pandemic prediction. The final part of the book title refers to the project that served as the basis for the realisation of this book.

In this chapter, the main ideas that guided the preparation of the book are presented. All concepts that are included in the admittedly long book title are

DOI: 10.1201/9781003148883-1

3

explained, and their integration into a unified framework justified. This will hopefully set the tone for the reader to understand what we did, why we did it, and the order they are introduced in the sequel. After this description, a summarised view of the following parts and chapters that constitute the book, and a guide with different suggested routes on how to read it are provided. The notation is also introduced and explained here.

1.1 Objective of the book

This book is about platforms for pandemic prediction. Therefore we must describe first what is meant here by a platform. Loosely speaking, it is any device that enables users to obtain information about a given matter. In our case, the matter of interest is pandemic prediction. This device usually comes in the form of an online application widely available on computers, notebooks, tablets, mobile phones and any other equipment with internet access. Online applications abound at this epoch over an immense variety of formats and purposes, and are usually referred to as apps. A more formal definition and a detailed description of platforms are provided in the sequel.

The book will go through the various steps involved in the preparation of such apps, with the purpose of prediction of relevant features of pandemics and epidemics. By prediction, we mean to provide a full description of the uncertainty associated with such a task. This would be the predictive distribution of the features of interest in the future, under the Bayesian paradigm for Statistics. The entire description of this object is the subject of later chapters. For now, it suffices to say that inference under this paradigm is entirely model-based and forecasts are always probabilistic. This means that inference relies entirely upon the construction of the probabilistic specification for observations and other quantities present in the model.

Prediction may be split into the categories short term and long term in many applications. There are no mathematical definitions for these terms, but they roughly refer to the immediate future and the distant future. This distinction is particularly pertinent in the context of pandemics. Short term for a pandemic means, in general, the immediate future of up to two weeks. Long term refers to the end of the pandemic or of its first wave, understood here as the first large mass of occurrences of the disease. This book addresses predictions for both terms. Both categories have their merits. The nature of the course of most epidemics through a human population is inevitably subject to change; conditions that were held in the past may no longer be valid after a while. Thus, short-term predictions are easier to perform well. The failure to provide good results for the long run have led to many criticisms by the general public but also in the scientific community (Ioannidis et al., 2020). Nevertheless, long-term prediction provides a useful indication of the magnitude of things to come (Holmdahl and Buckee, 2020). Rather than showing the way, the long-term predictions throw light into the way. They might provide useful indicators when used with caution. As such, they may constitute an important component towards a more encompassing view of the progression of a pandemic.

Pandemics occur worldwide, i.e., over hundreds of countries. Similarly, many epidemics occur over dozens of countries. A suitable prediction platform should aim at all countries involved in the epidemic or, at the very least, at a fair number of these countries. This brings in the unavoidable need for considering a large number of units for prediction. Also, this kind of problem deals with life-and-death situations. Instant updates of the prediction results after a new data release is imperative. In the case of COVID-19 and in many other epidemics, official data is updated daily. In the typical pandemic scenario, many analyses are required at high frequency, causing a considerable computational burden. The more elaborate and country specific the model is, the longer it will take to fit it to the data and to generate prediction results. The famous George Box motto of "all models are wrong but some are useful" (Box et al., 2005) could not find a more appropriate application than this. So, models should be carefully and parsimoniously chosen in order to include the most important features of the pandemic, but only them. This will probably not lead to the best prediction for every country, but hopefully will provide useful ones for most of them. In the sequel, a specific presentation of each of the main features of platforms for forecasting pandemics is given.

1.1.1 Data-driven vs model-driven

A relevant aspect of this book is the construction of the model for the observations based on the apparent features of the data. We refer to this approach as data-driven. This can be contrasted against model-driven approaches, that use biological/physical considerations to drive the model specification. This distinction is more didactic than practical, as there is frequently an interplay between the two sources of information. For example, there are many possible data-driven alternatives for model specification. They are mostly similar in qualitative terms. The ones highlighted in this book were based on some basic theoretical consideration. So there is a mild presence of theory in this choice. Therefore, purists might say that our approach is not data-driven or model-driven.

The data-driven approach fits in nicely with the pragmatic approach advocated in this book for real-time predictions. Having a parsimonious model as the baseline enables the incorporation of data features that may not be present at the onset of a pandemic, but may emerge at a later stage, as is the case with many pandemics, including COVID-19. The model needs to be enlarged in complexity to accommodate these features. But the computing time spared through the use of a simple starting model makes it feasible to allow for the inclusion of these additional features.

It must be acknowledged, however, that there are many studies of pandemics that are based on the model-driven approach. This literature, mostly based on the so-called compartmental models, known also as *SIR* and its extensions, has been considerably enlarged with the COVID-19 pandemic. The idea behind these models is to allocate the population into different compartments and to describe the dynamics of the population transitioning between them. The most basic version has three compartments, (S)usceptible, (I)nfectious and (R)ecovered.

The data-driven approach of this book is distinct from the various versions of compartmental models, even when the latter uses data to estimate their parameters. For this purpose, the basic features of the compartmental models are briefly described so that similarities and differences are pointed out, without getting into technical details.

The many versions of compartmental models have a few similar assumptions. They assume that the population under study is closed, that is, it does not suffer external influence. They also assume that the population is perfectly mixed, in the sense that every individual has the same (non-)contact pattern with every other individual. An additional assumption is that every susceptible person has the same probability of being infected. But there have been proposals trying to weaken these assumptions, see Grimm et al. (2021) for example. Then, the individuals are assigned to different compartments and the transition dynamics are described through ordinary differential equations. While the original SIR model was proposed in Kermack and McKendrick (1927), there are many extensions which add compartments such as (D)eceased (Parshani et al., 2010), (M)aternally derived immunity, (E)xposed (Hethcote, 2000), and others, as well as combinations thereof (Martianova et al., 2020).

One of the features of compartmental models is that they are deterministic, which means that, for a given parameter value, the dynamics between the compartments are completely defined and the progress of the disease is set. This is unrealistic as this ignores many uncertainties not covered in the differential equations, such as delayed case reporting, imperfect infection date recording and many others. To account for that, a probabilistic model can be added to aggregate uncertainty. Modern tools exist to estimate model parameters in this case, such as the pomp R package (King et al., 2016) and the LibBi library (Murray, 2015). Their use has been compared in Funk and King (2020).

Similar to the methodology presented in this book, compartmental models may use observed data to guide the choice of values for the parameters. The fundamental difference between any compartmental model and the methodology adopted here is that the former is built based on a description of the physical process behind the data, to which uncertainty could be added through a probabilistic model. The choice of how many and which compartments to add to the model is a major factor in the description of the physical process, fit to the data and parameters interpretation. In contrast, the proposal discussed throughout this book, although initially inspired by a growth model (details in Chapter 3), has its features and extensions entirely motivated by its data feature adequacy. Under no circumstances do the models presented in this book attempt to describe the physical process behind an epidemic or pandemic. This is an important distinction that warrants the data-driven 'stamp' as it is entirely and exclusively motivated by the features found in the data.

There are many other methods of mathematical modelling of epidemics, such as statistical regression techniques, complex networked models, web-based data-mining, and surveillance networks, among others. It is not the intention of this book to describe them, but a short summary can be found in Siettos and Russo (2013). A more in-depth and complete review can be seen in Held et al. (2019). An interesting model-based approach recently developed during the COVID-19 pandemic is implemented in the `epidemia` R package and can be read in detail in Scott et al. (2020).

1.1.2   Real-time prediction

In the emergence of an outbreak of a novel disease, or even when cases of a known malady start presenting an increasing pattern, projections of the most probable scenarios become remarkably important as countless fundamental decision-making processes depend on what to expect from the future (Fineberg and Wilson, 2009; Pell et al., 2018; Funk et al., 2019; Reich et al., 2019). The mentioned projections may concern what can happen in both short-term (days or weeks ahead) and long-term (months or the entire duration) periods. As mentioned in Funk et al. (2019), Viboud and Vespignani (2019), Roosa et al. (2020), and Chowell et al. (2020), short-term projections may serve as a guidance for authorities concerning prompt decisions such as organising hospital beds and individual protection equipment. In turn, long-term ones are related to a solid preparation of health care systems and vaccination programs, for example. However, the efforts concerning the mathematics behind these projections of future scenarios are not an easy task (Hsieh et al., 2010; Tariq et al., 2020). Viboud and Vespignani (2019) compare predicting cases of epidemics with the challenges involving weather forecasting, and competitions worldwide aim to encourage the improvement of models (see Pell et al., 2018). For instance, it can be the case that several factors (e.g., lockdown, quarantine, vaccination, among others) impose a rapidly change of the situation of the epidemic curve. Thus, as reinforced in Hsieh and Cheng (2006) and Chowell et al. (2020), projections may be obsolete in a short period of time, invalidating its practical usage. Consequently, the constant update of results, i.e., real-time predictions, may be even more crucial in such situations.

Real-time predictions depend on a constant update of data. This rigorous routine may impose a limitation on modelling approaches due to the lack of case-specific information such as gender, age, and disease-related dates, to cite a few (Pell et al., 2018). In addition to that, in extreme situations like a pandemic, data may have poor consistency since the gravity of the situation itself may prevent data from being carefully collected and computed (Funk et al., 2019). Data may also be reviewed (Hsieh et al., 2010). In the COVID-19 pandemic for instance, repositories like the Center for Systems Science and Engineering at Johns Hopkins University (Dong et al., 2020) displays basic –and important– information on counts of cases worldwide; but most of the data details mentioned above are not available.

Another important point to take into account is the evaluation of results. As extensively discussed in Funk et al. (2019), the use of comparison metrics on prediction outcomes, an appropriate quantification of uncertainty, and comparing modelling alternatives assume central roles in reaching a final goal of providing good information.

The conclusion of this discussion is that modelling this type of data aiming at (at least) providing real-time predictions requires routines that are: a) complex enough to provide coherent results, but b) parsimonious to allow for timely production of outcomes. This way, updated projections can be used routinely for planning and evaluating strategies.

1.1.3   Building platforms

After obtaining the predictions, the next step is to present the results in the best possible format. There are cases where these results are only required for utilisation by a single person, a small group of people or an individual institution. In these cases, there might not be any need for a structured platform. In most cases, however, an application considering all the information that needs to be conveyed, as well as the different users' profiles, must be constructed. The recipient public can range from other statisticians and academic researchers, government officials, to journalists and the general public, accessing the platform from all over the world. The application should allow them to visualise the most recent data, in addition to the updated prediction results for a set of different input options. When choosing how to display the data, it is important to consider the adequacy to different devices that can be used to access the application, as well as the variety of people's backgrounds and their graph literacy.

The plots' aesthetics are crucial to clearly display the observed values, the predictions and their uncertainty, and any other relevant model components. Interactive plots can help to select regions to zoom into, identify specific values and choose which data series to show. Another feature that can be beneficial for public use is the option to download the plots and data files. These are some of the advantages of publishing the results in an online interactive application.

Some examples of nice and useful platforms that provide a great amount of data displayed in various formats can be seen in the website of the Our World in Data organisation (`ourworldindata.org`), that among other topics, has a designated section for visualisation of data about the COVID-19 pandemic. The Institute for Health Metrics and Evaluation (IHME) also has a variety of applications designed to disclose global health research results in different formats. It is worth highlighting their platform with worldwide projections related to the COVID-19 pandemic available at covid19.healthdata.org/projections. These are just some examples of the diversity of these data visualisation platforms. There are actually no limits on how complex the platform functionalities can be. This will depend mostly on the availability of human and computational resources, but it is also related to the objective of each project.

Our choice for the language and tools to build such an application was R (R Core Team, 2020) and the Shiny package (Chang et al., 2020), since our team consisted mainly of statisticians. This package allowed us to easily create and publish a user interface to select the country/region of choice and readily see the latest data and predictions. The user interaction with the plots was also available thanks to the plotly graphing library (Sievert, 2020). We believe that these might be the choices of other researchers as well, such as statisticians and epidemiologists. They will probably follow similar paths in the process of publishing some model fit results. That is why the main steps and issues of our process in building such a platform are compiled in this book, in the hope that we can help others in their own journeys.

It should be noted that the application can be part of a broader integrated platform according to the projects' needs. In the case of the CovidLP Project, a site to include more methodological details, announcements and updates related to the project, and a channel for user interaction was also developed. Besides this, all the source code and latest results are openly available in an online repository, enabling collaboration among more advanced users. Lastly, the code used to fit the model and obtain the predictions was turned into an R package to facilitate replication of the methodology to other data sets.

Either way, the book succinctly covers all the material (models, platform and software) presented above. But the development of the platform, our assumed goal, goes way beyond model specification and its ensuing inference. It also goes beyond the construction of a computing software.

Some of the following chapters elucidate the procedures needed to construct such an integrated platform, starting from data extraction, model selection and fit, to publishing the prediction results. A concern common to all these steps is the automation of processes, especially when dealing with dynamic data sets such as the ones for real-time pandemic prediction. It is important to think about how to automate the whole data-driven process when wanting to provide the most up-to-date predictions for a large number of countries and regions. This scenario is determinant for many of the decisions taken in this book. These issues will be hopefully clarified in the respective parts of the book, to be described in the next section.

1.2   Outline of the book

This book is divided into seven parts, each one of them with a number of chapters. Part I contains 2 chapters. Chapter 1 provides an introduction to the book by setting the scene that governs our approach, as described in the previous section, and this description of the content of each chapter. Chapter 2 deals with a discussion of the basic input for pandemic prediction: the data of a pandemic. The main data output of pandemics and epidemics (confirmed cases and death counts) is described and then discussed. Other data sources that might be relevant for the understanding of the process are also presented, discussed and compared, and their relation established with primary data sources.

Part II introduces statistical modelling. It is divided into four chapters for ease of presentation and understanding. Chapter 3 presents and discusses the main epidemiological inputs of the temporal evolution of the pandemic that led to the (generalised) logistic form. Alternative formulations are also provided. Properties that are useful for understanding and for communicating the main features of pandemics/epidemics are presented and evaluated for these specifications. Chapter 4 presents possible data distributions for pandemic data, starting from the canonical Poisson specification. Overdispersion is a relevant concept and different forms to introduce it are presented and compared. Once the probabilistic specification and data distributions are set, the model for the observations is completely specified and inference can be performed. Chapter 5 presents many other data features that appear in some observational units due to their departure from the basic hypothesised model. Finally, Chapter 6 presents a review of the Bayesian inference approach used in this book. Prior distributions will play an important stabilising role, especially when scarce data is available, e.g., at the early stages of the pandemic. Approximations are required in cases when analytical results are not available, as is the case with most results of this book. The techniques used for approximations are also briefly presented.

Part II presents the basic ingredients for modelling but does not exhaust the many advances that were proposed in the literature. It only describes the main pandemic features, directly observed from the data. They can be modelled in a single layer specification, without the need for further structure.

Part III deals with situations where extra layers are required for appropriate modelling. These additional layers or levels are required to accommodate variations from the basic structure. Chapter 7 handles statistical solutions to many of the data problems that are identified in Chapter 2. One of the most important problems is under-reporting, a prominent issue in the data analysis of many pandemics/epidemics. Another relevant issue is that pandemic counts are based on the false assumption of perfect identification of cases (and deaths). Approaches to address the above issues are presented. Chapter

8 presents a number of alternative models that address relevant extensions of the modelling tools of Part II. They are based on the hierarchical specification of the model with the presence of additional layers. These layers introduce temporal, spatial and/or unstructured dependence among units or times.

Part IV deals with the implementation of these ideas in a platform to display the prediction results. Chapter 9 outlines the procedure of automating data extraction and preparation. This ETL (Extract, Transform and Load) stage is essential when dealing with different data sources for several countries that are constantly being updated. Chapter 10 explains the steps for automating modelling and inference, considering scalability and reproducibility. A review of the Bayesian software options available and their characteristics is also presented. Lastly, Chapter 11 describes the development of the online application. A brief tutorial on how to build similar platforms with `Shiny` is presented, as well as details about all of its elements including the interactive plots to display the forecasts.

Every statistical analysis should be subject to scrutiny. Part V handles this task in its various configurations. Chapter 12 describes data monitoring schemes to anticipate their possible departure from an underlying trend. Chapter 13 describes how to set up a constant monitoring scheme on a routine basis for early identification of depreciation of prediction performance. Chapter 14 provides comparison results against similar platforms for pandemic prediction.

Part VI describes the R package `PandemicLP`, the software developed by our team to perform the analyses and produce the predictions. Chapter 15 provides an overview of the package basic functionalities, and presents examples to help the users to apply our proposed methodology to their own data sets. Chapter 16 addresses more advanced features already available in the software. These include settings to control MCMC efficiency and tools for sampling diagnostics.

The book is concluded with Part VII. There, a single chapter briefly addresses some of the relevant points that the future editions of this (or a similar) book could cover. It discusses some of the possibilities for extensions to the framework of the book in all its dimensions: modelling, implementation, monitoring and software.

The book also has its own supplementary material repository. It is located at `github.com/CovidLP/book`. It contains files related to the book such as code and databases for the figures and tables of the book. It is worth mentioning that the book repository above is one of a number of repositories from the CovidLP project, freely available at the encompassing repository `github.com/CovidLP`.

1.2.1 How to read this book

This book is unavoidably interdisciplinary, especially in terms of Statistics and Computing. Therefore, it may serve different purposes for readers with different backgrounds. We describe next some possible ways to read this book depending on the type of information the reader is seeking.

First, the trivial path is to read the seven book parts in sequence, which may serve users interested in learning about all the stages of our platform-building process, focused on the COVID-19 pandemic forecasting. This path is also instructive to whomever wants to replicate the entire process of our project, with similar modelling frameworks.

Another possible reading path is geared to users interested mainly in the methodological aspects of modelling epidemic data. For those users, we recommend reading Part I for introduction and data description, and Parts II and III for basic and further modelling aspects, respectively. Included at the end of Part II is Chapter 6, which presents a review of Bayesian inference. This chapter is left as an option for the reader who might feel the need to learn/revisit the main concepts used within the book. The user following along this reading path might also be interested somewhat in Part VI, where the implemented functions from the R package to fit the proposed models are exemplified.

For users searching for instructions about how to create and maintain an online platform for up-to-date presentation of some statistical results, we suggest following Part I with the reading of Parts IV and V. These parts include the step-by-step handbook of how to create an online application with automatic data extraction and obtaining of predictions, in addition to the discussion of some important features to monitor on inference results.

The book is finalised in Part VII, where most parts are revisited with an introductory, concise view to summarise possible directions for the future. This part might be of interest to all readers not satisfied just by what was done, but also on what could come next.

1.2.2 Notation

Most pandemic data consist of counts, usually counts of infected cases or deaths caused by the disease. The counts can be recorded separately for each time unit or recorded by accumulation over the previous time units. The latter is the result of integration (or sum) of the former. The usual mathematical standard is to use capital letters for the integrated feature and small case for the integrand. Thus, cumulative counts will be denoted by Y, while their counts over a given time unit will be denoted by y.

Counts are collected over periods of time, that could be days, weeks, months, etc. Whatever the time unit, the counts at time t are denoted by y_t, while the cumulative counts up to time t are given by $y_1 + \cdots + y_t$ and denoted by Y_t, i.e., $Y_t = \sum_{j=1}^{t} y_j$, for all t.

Typically, these counts are random variables with finite expectations or means. In line with the distinction between cumulative and time-specific counts defined above, the means are denoted by $M(t) = E(Y_t)$ for the cumulative means, and $\mu(t) = E(y_t)$ for the time-specific means, i.e., $M(t) = \sum_{j=1}^{t} \mu(j)$, for all t. The dependence on time is denoted for the means in the most usual functional form because their dependence on time will be made explicit, unlike the counts that will depend on time implicitly. These points will be made clear in Part II.

Bibliography

Box, G., Hunter, J. and Hunter, W. (2005) *Statistics for Experimenters: Design, Innovation, and Discovery.* Wiley Series in Probability and Statistics. Wiley. URLhttps://books.google.com.br/books?id=oYUpAQAAMAAJ.

Chang, W., Cheng, J., Allaire, J., Xie, Y. and McPherson, J. (2020) *shiny: Web Application Framework for R.* URLhttps://CRAN.R-project.org/package=shiny. R package version 1.5.0.

Chowell, G., Luo, R., Sun, K., Roosa, K., Tariq, A. and Viboud, C. (2020) Real-time forecasting of epidemic trajectories using computational dynamic ensembles. *Epidemics*, **30**, 100379.

Dong, E., Du, H. and Gardner, L. (2020) An interactive web-based dashboard to track COVID-19 in real time. *The Lancet Infectious Diseases*, **20**, 533–534.

Fineberg, H. V. and Wilson, M. E. (2009) Epidemic science in real time. *Science*, **324**, 987–987.

Funk, S., Camacho, A., Kucharski, A. J., Lowe, R., Eggo, R. M. and Edmunds, W. J. (2019) Assessing the performance of real-time epidemic forecasts: A case study of Ebola in the Western Area region of Sierra Leone, 2014-15. *PLOS Computational Biology*, **15**.

Funk, S. and King, A. A. (2020) Choices and trade-offs in inference with infectious disease models. *Epidemics*, **30**, 100383.

Grimm, V., Mengel, F. and Schmidt, M. (2021) Extensions of the SEIR model for the analysis of tailored social distancing and tracing approaches to cope with COVID-19. *Scientific Reports*, **11**.

Held, L., Hens, N., O'Neill, P. and Wallinga, J. (eds.) (2019) *Handbook of Infectious Disease Data Analysis.* Boca Raton: Chapman and Hall/CRC, 1st edn.

Hethcote, H. W. (2000) The mathematics of infectious diseases. *SIAM Review*, **42**, 599–653.

Holmdahl, I. and Buckee, C. (2020) Wrong but useful – What Covid-19 epidemiologic models can and cannot tell us. *New England Journal of Medicine*, **383**, 303–305. URLhttps://doi.org/10.1056/NEJMp2016822.

Hsieh, Y.-H. and Cheng, Y.-S. (2006) Real-time forecast of multiphase outbreak. *Emerging Infectious Diseases*, **12**, 122–127.

Hsieh, Y.-H., Fisman, D. N. and Wu, J. (2010) On epidemic modeling in real time: An application to the 2009 novel A (H1N1) influenza outbreak in Canada. *BMC Res Notes*, **3**.

Ioannidis, J. P., Cripps, S. and Tanner, M. A. (2020) Forecasting for COVID-19 has failed. *International Journal of Forecasting*. URLhttp://www.sciencedirect.com/science/article/pii/S0169207020301199.

Kermack, W. O. and McKendrick, A. G. (1927) A contribution to the mathematical theory of epidemics. *Proceedings of the Royal Society A*, **115**, 700–721.

King, A. A., Nguyen, D. and Ionides, E. L. (2016) Statistical inference for partially observed Markov processes via the R package pomp. *Journal of Statistical Software, Articles*, **69**, 1–43.

Martianova, A., Kuznetsova, V. and Azhmukhamedov, I. (2020) Mathematical model of the COVID-19 epidemic. In *Proceedings of the Research Technologies of Pandemic Coronavirus Impact (RTCOV 2020)*, 63–67. Atlantis Press.

Murray, L. M. (2015) Bayesian state-space modelling on high-performance hardware using LibBi. *Journal of Statistical Software, Articles*, **67**, 1–36.

Parshani, R., Carmi, S. and Havlin, S. (2010) Epidemic threshold for the susceptible-infectious-susceptible model on random networks. *Phys. Rev. Lett.*, **104**, 258701.

Pell, B., Kuang, Y., Viboud, C. and Chowell, G. (2018) Using phenomenological models for forecasting the 2015 Ebola challenge. *Epidemics*, **22**, 62–70. The RAPIDD Ebola Forecasting Challenge.

R Core Team (2020) *R: A Language and Environment for Statistical Computing*. R Foundation for Statistical Computing, Vienna, Austria. URLhttps://www.R-project.org/.

Reich, N. G., Brooks, L. C., Fox, S. J., Kandula, S., McGowan, C. J., Moore, E., Osthus, D., Ray, E. L., Tushar, A., Yamana, T. K., Biggerstaff, M., Johansson, M. A., Rosenfeld, R. and Shaman, J. (2019) A collaborative multiyear, multimodel assessment of seasonal influenza forecasting in the United States. *Proceedings of the National Academy of Sciences*, **116**, 3146–3154.

Roosa, K., Lee, Y., Luo, R., Kirpich, A., Rothenberg, R., Hyman, J., Yan, P. and Chowell, G. (2020) Real-time forecasts of the COVID-19 epidemic in China from February 5th to February 24th, 2020. *Infectious Disease Modelling*, **5**, 256–263.

Scott, J. A., Gandy, A., Mishra, S., Unwin, J., Flaxman, S. and Bhatt, S. (2020) epidemia: Modeling of epidemics using hierarchical Bayesian models. URLhttps://imperialcollegelondon.github.io/epidemia/. R package version 0.7.0.

Siettos, C. I. and Russo, L. (2013) Mathematical modeling of infectious disease dynamics. *Virulence*, **4**, 295–306.

Sievert, C. (2020) *Interactive Web-Based Data Visualization with R, plotly, and shiny.* Chapman and Hall/CRC. URLhttps://plotly-r.com.

Tariq, A., Lee, Y., Roosa, K., Blumberg, S., Yan, P., Ma, S. and Chowell, G. (2020) Real-time monitoring the transmission potential of COVID-19 in Singapore, March 2020. *BMC Medicine*, **18**.

Viboud, C. and Vespignani, A. (2019) The future of influenza forecasts. *Proceedings of the National Academy of Sciences*, **116**, 2802–2804.

2

Pandemic data

Dani Gamerman

Universidade Federal de Minas Gerais/Universidade Federal do Rio de Janeiro, Brazil

Vinícius D. Mayrink

Universidade Federal de Minas Gerais, Brazil

Leonardo S. Bastos

Fundação Oswaldo Cruz, Brazil

CONTENTS

Data is the primary input for any statistical analysis. In this chapter we present the main aspects of pandemic data, starting from their definitions. Their virtues and deficiencies are described and compared. We focus on data used for and in the predictions. Auxiliary variables that are related to the primary data variables are also described and their relations are presented.

2.1 Basic definitions

It is first important to establish exactly what is being predicted in order to make pandemic prediction. Clear definitions are required from the beginning. An obvious pre-requisite is to know what a pandemic is. According to Last (2001) a pandemic is "an epidemic occurring worldwide, or over a very wide area, crossing international boundaries, and usually affecting a large number of people". This definition calls for the definition of epidemic. The same dictionary informs us that an epidemic is "the occurrence in a community or

DOI: 10.1201/9781003148883-2

region of cases of an illness, specific health-related behaviour, or other health-related events clearly in excess of normal expectancy". These definitions are quite general including epidemics of communicable and non-communicable diseases, risk factors like obesity, or even the spread of information like fake news. It is worth mentioning that our primary concern is to deal with epi/pandemic of infectious or communicable diseases, like the coronavirus infectious disease (COVID-19) caused by the coronavirus SARS-CoV-2.

A few comments are in order here. The first comment is that the definitions above are relative in the sense that their very definition is comparative against a "normal" situation. The second comment is that the definitions above do not account for severity or other disease characteristics. This is possibly to avoid more subjectivism and hence to avoid criticisms over the definitions. Nevertheless, this cautionary approach did not prevent authors from questioning them (Doshi, 2011). We will avoid this discussion hereafter and assume that the diseases we will be predicting meet the epi/pandemic criteria.

The third comment has a deeper impact on this book and deserves a full paragraph. The definitions above make it clear that the major difference between a pandemic and an epidemic is its reach. When an epidemic affects a large number of countries or the entire world, as for example happened with the COVID-19 in 2020, then it becomes a pandemic. Therefore, if one is considering predicting only a single geographical region (be it a continent, a country, a state or a city) without taking into account other regions, the same statistical models and methods can be indistinguishably used for either a pandemic or an epidemic.

There are a number of basic statistics associated with any disease. By far, the most used ones are the counts of cases and deaths of the disease. They appear as the major statistics from bulletins that are routinely issued by health authorities worldwide, especially during the course of an epi/pandemic. Once again, the concepts of cases and deaths of a disease must be clearly defined. These are usually defined based on clinical and/or laboratory criteria. Note that these criteria may change across regions or, even worse, over time during the course of a pandemic. Changes across regions make it difficult to compare them as the counts may be reporting different characteristics. Changes across time make it difficult to associate a single temporal pattern and to associate a relation between successive times, complicating the statistical analyses of their evolution.

It is crucial to have counts of a disease in a timely fashion. They help health officials make decisions on allocation of resources (personnel and equipment) and let the general public be aware of the current situation and its evolution. They also allow the immediate update of prediction systems, providing further information to the society. The relevance of fast disclosure of the data available obviously depends on the severity of a disease. Most recent epidemics and pandemics are undoubtedly severe enough to give an urgency to the need for the release of information. This urgency has an important impact over the data protocols for data disclosure. Throughout the world, they typically involve the daily release of data collected up to that day. These are usually referred to as confirmed cases of a disease.

Other time windows are also obtained for some pandemics. By far the second most common time frame for data release is weekly data, especially for diseases with very low counts. This also allows for removal of possible weekday effects that exist in some health notification systems. Another strategy commonly applied in diseases with weekday variations is the use of 7-day moving averages to smooth data counts.

The distinction between different categories of cases will be returned to in the next section. Figure 2.1 illustrates the data mentioned above for the COVID-19 pandemic for Switzerland.

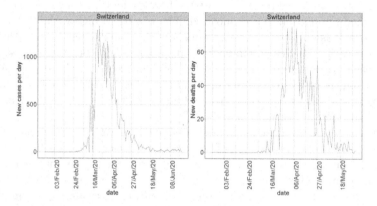

FIGURE 2.1: Daily data of confirmed cases and confirmed deaths of COVID-19 for Switzerland.

Figure 2.1 also includes the number of confirmed counts associated with deaths caused by the disease. One striking feature of the figure is the qualitative similarity of the curves of confirmed cases and confirmed deaths observed for the same region. This similarity occurs despite the substantial difference in nature of the two types of counts. This similarity is observed in many other countries across the globe and will be explored by the data-driven approach throughout the book.

In the context of a infectious disease epidemic, some individuals may be infected without manifesting any symptom. These individuals, however, can still transmit the agent that causes the disease. For example, a person infected with the human immunodeficiency virus (HIV), called HIV-positive, may be asymptomatic for several years. Without adequate treatment, a person with HIV can develop the disease AIDS (acquired immune deficiency syndrome).

In any case, diseases manifest themselves through their symptoms. And we just noted above that some infected individuals may be asymptomatic in the sense they may not manifest any symptom and may carry on their normal life totally unnoticed. Nevertheless, even though harmless to these individuals, they may infect other people. Some of the newly infected ones may well present serious symptoms and the disease may even evolve to their deaths. Thus, asymptomatic cases are just as important as the symptomatic cases but they are typically harder to identify. If the symptoms do not show up in an individual, it becomes very hard to identify him/her as a case. The only way it could happen is through a widespread testing campaign in the region of the individual. These campaigns are costly and present many logistics difficulties.

The COVID-19 disease is a timely and important illustrative example. Identification of cases is typically achieved through molecular tests such as RT-PCR (reverse transcription polymerase chain reaction). These tests are not widely available in many countries and even when they are, they may take days to have their results released. These difficulties cause large parts of the infected population to go unnoticed. There is still a high variability for the proportion of asymptomatic cases among studies. Kronbichler et al. (2020) was one of the earlier studies and indicated this proportion to be around 62%. So, a substantial proportion of under-reporting is to be expected unless population-wide testing is performed. Tests are also imperfect, even molecular tests are due to error leading to false-positive and false-negative results. Imperfect tests will be revisited in Section 7.3.2.

Data for diseases must be handled with care for the above reasons. When there is a substantial proportion of asymptomatic cases, counts will be affected by this feature. Therefore, it is inappropriate to refer to the obtained counts as total counts. This issue is clarified with the nomenclature used for data released on pandemics. Instead of referring to these measurements simply as counts, they are named confirmed counts. This standard is widely used and will be considered throughout this book.

2.2   Occurrence and notification times

As mentioned in the previous section, pandemic data released daily by health authorities present counts that are compiled according to individuals that become known to be cases (or deaths) at a given day. These counts will be referred to as counts by notification date and these are the confirmed counts, referred to in previous section. This association of cases with the date they are notified is the simplest form to compile pandemic data, but it is far from perfect. Cases are notified at the date they become publicly available by the health authorities. It is well known that there is an inevitable delay between the date a case occurs and the date the health system becomes aware of this event, after being notified. This delay can range from hours to weeks, depending on the efficiency of the health data collection system. Clearly, the mentioned delay has an important impact on the counts.

Counting by notification date is clearly simpler, but it has important methodological drawbacks. Assume a case is notified a week after it actually occurred. If this case is only added to the counts at its notification date, it will provide delayed information about the evolution of the disease, biasing the inference. The magnitude of the bias may be noticeable if this delay impacts a substantial proportion of the cases. A qualitatively similar argument is valid for counting deaths. Example 2.1 below provides a numerical illustration of the problem.

Example 2.1: Let 100 cases be notified on day X. Assume that the occurrence dates are distributed as follows:

Occurrence day	X	X-1	X-2	X-3	X-4	X-5	X-6	X-7	X-8	X-9
	55	15	12	8	4	1	0	2	2	1

Thus, only 55 of these cases actually occurred at day X. All the other 45 individuals became cases at an earlier day.

Therefore, it seems more appropriate to monitor any pandemic through cases counted by occurrence date. In fact, there is little doubt about the adequacy of this choice. Why is this protocol not used worldwide? One possible explanation is the information delay occurring even for regions with entirely organised health databases. If counts arrive to the health information system with a delay, then total counts that occurred at a given date would not be available on the same day. Therefore, if a real-time analysis is needed, the most up-to-date counts are poorly represented. So in order to use the occurrence date, it is necessary to correct the delay. This issue will be explored in more detail in Section 7.2.

Another problem related to occurrence dates is the difficulty in obtaining such data. Ascertainment of the exact occurrence date requires going back to each individual file. This task is far from trivial as typically the data are obtained at a disaggregated level (district or city) and informed at an aggregated level. This would require an integrated system as these operations must be registered digitally for fast dissemination of information.

Example 2.1 (continued): Assume that further cases, which occurred at day X, were notified at the following days:

Notification day	X	X+1	X+2	X+3	X+4	X+5	X+6	X+7
	55	20	8	4	2	0	0	1

So, the further 35 cases actually occurred at day X, but were reported at a later date. Therefore, the number of cases that occurred at day X was actually 90, while the number of cases that were notified at day X is 100. The number of cases known at day X, that occur at that day, is just 55. So, using the counts by occurrence date, it is necessary to estimate the cases that occurred, but have not been reported yet using a statistical method.

2.3   Other relevant pandemic data

The count data expressing the number of confirmed cases or the number of deaths are not the only type of relevant information that might be considered to study a pandemic. Some key conclusions can also be obtained from variables such as hospital bed and ICU (intensive care units) occupancies. In many regions, these elements are constantly updated and reported by local health authorities during the pandemic. This is done to inform the public about the current situation of the health system and also to indicate results that justify government actions to prevent the disease from spreading in the population.

Several studies are dedicated to exploring hospital capacity data in different countries. Some few examples, in the COVID-19 context, are Grasselli et al. (2020), Moghadas et al. (2020), Rainisch et al. (2020) and Salje et al. (2020). The demand for hospitalisation of COVID-19 patients is affected by different factors such as preexisting health conditions and age. The impact of this demand on the society is expected to change as the pandemic advances and the knowledge about the disease improves. Projections of hospital occupancy are critical for planning purposes so that the health system can gather resources in advance of the high demand. Some aspects to be considered for occupancy projections are: number of available beds, non-pandemic occupancy rates, epidemiological curve of the disease and length of stay for affected patients.

The number of recovered patients is another variable that might be explored in a study to understand the pandemic progress. However, in an outbreak such as COVID-19, the recovery counts are almost always inaccurate, since the observed values are potentially lower than the true ones. One major reason for this is the fact that most affected individuals are mild or asymptomatic cases, therefore, they will not be included in the official reports if the testing capacity is low in the region where they live. In addition, people tend not to report back to update the health care provider when they are recovered from a disease. A strategy to improve the accuracy of the recovery counts is to follow up with all the patients to verify their conditions, but this is too costly when compared to other public policies to handle the pandemic. Even though the recovery counts are problematic, many statistical analyses are based on this type of variable; some examples are: Anastassopoulou et al. (2020), Chintalapudi et al. (2020), Yadav et al. (2020) and AL-Rousan and AL-Najjar (2020).

The number of recovered patients can be aggregated with the number of deceased patients to form what is usually referred to as the number of removed patients. This count totals the number of individuals that were infected and, for entirely opposite reasons, are no longer able to infect other individuals. This number is an important property in epidemiological studies, as it represents the number of individuals that are still infecting others. This point will be returned to in Chapter 3.

The level of testing to detect the infected individuals in a population is a useful variable to explain the pandemic progress. The true total number of infected people is clearly unknown. The known infection status available from official data repositories is related to individuals who have been tested. This implies that the counts of confirmed cases depend on how often the detection tests are applied in a region. In order to control and properly monitor the spread of the disease, increasing the number of tests is critical. A high testing level provides more reliable data. An interesting metric accounting for the testing level relative to the outbreak size is the so-called positive rate. A small result indicates that the pandemic is considered under control; in the COVID-19 context, the threshold is 5% according to the World Health Organization (WHO). For a given region, an increase in the positive rate across time is an indication that the disease is spreading faster than the growth of the confirmed cases curve. Alternatively, one can also evaluate the testing level based on the number of tests required to detect one infected individual. The WHO suggests 10 to 30 tests per confirmed case as a reference for an appropriate testing level. It is important to highlight that the variable "number of tests" may not represent the same thing in different data sets. In some countries this may include the same individual being tested more than once. Some few references accounting for testing levels of COVID-19 are Manski and Molinari (2020), Omori et al. (2020) and Tromberg et al. (2020).

The spatial dissemination of an infectious disease is clearly affected by the population movements within and between regions. Monitoring human mobility is an important point to be considered in epidemic evaluations. A central aspect for public health authorities here is to identify whether implemented social distancing policies are effective, and thus adjust them in locations where people do not comply with the rules. Public health planning will be more efficient if reliable information about human mobility is available for analysis. Some references in the literature indicate that mobile phone data can be used to capture the mobility level and to understand the geographical spread of an infectious disease; see Bengtsson et al. (2015) and Wesolowski et al. (2015). During the COVID-19 pandemic and other recent epidemics, governments and researchers, in different countries, have established collaborative efforts with mobile network operators and location tracker companies to access records providing individual geo-temporal information. The quantified human mobility is a key component for a spatial-transmission analysis of an infectious disease such as the COVID-19. Some additional references about using human mobility data are Oliver et al. (2020) and Tizzoni et al. (2014).

The measurements known as morbidity and mortality describe the progression and severity of a given disease in a population. They are useful to learn about risk factors and to compare the disease impact between different populations. Morbidity rate is commonly obtained using prevalence ($Y_t/N_{\text{Pop.}}$) or incidence ($y_t/N_{\text{Pop.}}$); where, for a particular period, $N_{\text{Pop.}}$ is the population size, Y_t is the total number of cases at time t, and y_t is the number of cases at time t. Mortality is based on the number of deaths caused by the target disease. The mortality rate is usually given by a rate per 100,000 individuals, having the same expression for morbidity with cases replaced by deaths. The mentioned quantities are based on retrospective information. Thus, they can be explored in a sequential analysis to evaluate the status of the health care system or the performance of a public health policy. In a pandemic scenario, an interesting study is to compare the morbidity and mortality rates of a region with the corresponding results from previous years. The excess of a morbidity or mortality rate from consecutive years is a useful indicator to evaluate eventual pattern alterations caused by the pandemic. Note that the comparison involving morbidity or mortality rates can also be done between groups of individuals with different characteristics. Some works related to the analysis of morbidity/mortality rates are Dudley and Lee (2020), Laurens Holmes et al. (2020), Jin et al. (2020) and Obukhov et al. (2020).

2.4   Data reconstruction

By data reconstruction we mean a set of statistical tools to correct common problems related to official epi/pandemic data such as reporting delay, undernotification and imperfect testing. In this section, these data problems are described and in Chapter 7 some existing methods to rebuild the data are presented and discussed.

Official disease counts, as previously mentioned, may be presented as occurrence or notification times. The number of cases by occurrence time shows the actual number of cases occurred in each day. These numbers are constantly updated as cases that occurred, but have not being reported yet. They are recorded at a later date in the official surveillance systems. On the other hand, the number of cases by notification date represents all new cases that entered the surveillance system regarding when they occurred. These numbers are not retrospectively updated. The difference between number of cases by notification and occurrence dates is illustrated in Figure 2.2, where new hospitalised COVID-19 cases in Brazil are presented by notification date and also occurrence date. Note that new cases reported by notification date is consistently smaller in weekends and holidays. This feature is observed because cases are notified by people that usually take these days off. One common correction is to use moving averages. The number of cases by occurrence date does not suffer from this drop of case reporting on weekends and holidays. However, the number of cases in the past must be daily updated as also indicated in Figure 2.2. This is caused by reporting delays.

The concept of reporting delays was illustrated in Example 2.1. Suppose today is day X. Then, assume that 90 cases are known to have occurred at

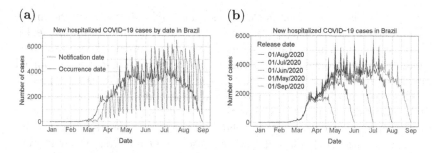

FIGURE 2.2: Daily hospitalised COVID-19 cases in Brazil: (a) by occurrence and notification dates; (b) using date of occurrence by different dates of data release.

day X, but only 55 were reported at that day. The remaining 35 cases were reported later with some delay. In practice, these delays are very common in epidemic data, and they may happen due to several reasons, for example, infrastructure problems in the health unit, lack of trained people to report the cases to the surveillance team that organises all data together before release, lack of tests if a disease case is only confirmed with a laboratory test, among many other reasons.

During an epidemic, case and death counts of the associated infectious disease, reported by occurrence date, may be incomplete because the cases and deaths are due to reporting delay. Suppose a new case is identified in a health unit; this case must be reported into a surveillance system. The local surveillance team will double-check the case, and this case will be added to the total of new cases for the next release. This process can take days, sometimes even weeks. For example, in Figure 2.2 the number of hospitalised cases of COVID-19 in Brazil by occurrence date with data released in early September by the end of August is nearly zero. This fast reduction in the number of new cases is more likely to be related to reporting delays than to a genuine reduction in the number of hospitalised COVID-19 cases in Brazil. In order to illustrate the delay problem, it can be seen from Figure 2.2 that on the 01 May, only three hospitalised cases of COVID-19 were reported that day. However, on 01 June, there were 2,304 hospitalised cases reported at 01 May, and by the end of September, there were 3,127 hospitalised cases reported as new cases in 01 May. These massive differences are due to reporting delays. If looking at cumulative cases until May, one can see 28,168 in May, and by September, these numbers were then updated to 57,890 cases.

The reporting delay issue is a well-known problem in the analysis of surveillance data. Usually, the surveillance teams that analyse the official data ignore the last K weeks of data, where K depends on the definition of the case that could be, for instance, a case informed by a health professional or a case confirmed with a laboratory test. However, by ignoring the most recent observations, the surveillance team may fail to identify in advance, for example, an unexpected rise in the number of cases that would characterise the initial phase of an outbreak. Had they known it in advance, additional public health measures could have been taken to mitigate the outbreak. There are some methods used to reconstruct, or estimate the occurred-but-not-yet-reported counts. Such methods are called "nowcasting", since they are used to predict what happens "now". These techniques will be introduced in Chapter 6.

It is important to point out that there is a difference between the unreported cases related to notification delay and those not reported due to other reasons. As mentioned before, unreported cases due to delay issues will eventually be reported. On the other hand, cases unreported for other reasons may not ever be reported, which is referred here as under-notification. For instance, some health units may not be prepared to apply laboratory tests. So a disease like COVID-19, that requires a laboratory confirmation, may be under-reported in that health unit. There are some methods that aim to reconstruct under-reported data using auxiliary data, like poverty indicators for instance.

As previously mentioned, some infectious diseases can only be confirmed after a test from a biological sample taken from the individual. For example, a person can only be confirmed to be infected with the HIV virus, if a HIV test confirms the infection status. Usually, such tests are performed in a laboratory, so a biological sample should be taken from the patient and sent for analysis. Afterwards, the result is sent to the patient or his medical doctor. The case is eventually notified to public health authorities, and the whole process may take days, or even weeks, leading to reporting delays. Laboratory delays must be taken into account in a surveillance system that intends to monitor an epidemic. A common solution that speeds up the process is the use of rapid tests, where the result is known in a couple of hours or sometimes minutes. Rapid tests are normally used for screening, where a patient with a positive result is strongly recommended to do a retest to confirm the disease status. From an epidemiological point of view, rapid tests are useful for population-based studies aiming to estimate a disease prevalence. For example, during the COVID-19 pandemic, Spain and Brazil have performed nationwide seroprevalence studies using rapid tests attempting to learn the extent of the epidemic in each country. See Pollán et al. (2020) for the Spanish study and Hallal et al. (2020) for the Brazilian one.

However, it is worth emphasising that no test is perfect. They are all prone to errors, i.e., they can return false results. The quality of a test is given by two probabilities known as sensitivity and specificity. Sensitivity of a test is the probability of a positive result given an infected person is being tested. The specificity, on the other hand, is the probability of a negative result when a non-infected person is tested. Hence, by knowing the specificity and sensitivity of a test, prevalence or incidence estimates, where the cases are defined using imperfect tests, can be corrected using probability rules.

External factors also affect the sensitivity and specificity of a test, for example: a biological sample may be taken outside the detection window, i.e., too early or too late to detect the pathogen; the sample could be damaged during the transportation or manipulation before the test itself; untrained professional running the tests could also be an issue; etc. All these mentioned possibilities may lead to false negatives, which implies a smaller sensitivity than the sensitivity suggested in optimal laboratory conditions. Figure 2.3 presents all hospitalised cases in Brazil, reported up to 31 August 2020, with the severe acute respiratory illness (SARI) that have been tested for SARS-CoV-2 using quantitative real-time reverse transcriptase-polymerase chain reaction (RT-PCR) assay. More than half a million tests were run and 260,000 of them were positive for SARS-CoV-2, therefore, hospitalised COVID-19 cases. In addition, 262,000 test results were negative for SARS-CoV-2. Does this result mean that there is an unknown pathogen causing hospitalised SARI cases? Most certainly not, and there is a certain proportion of these negative results that are in fact COVID-19 cases. Reconstruction of the actual number of hospitalised COVID-19 cases is important to understand the disease dynamic and estimate its extent appropriately.

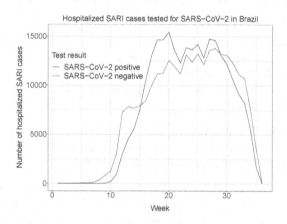

FIGURE 2.3: Weekly hospitalised SARI cases in Brazil tested for the SARS-CoV-2.

Bibliography

AL-Rousan, N. and AL-Najjar, H. (2020) Data analysis of coronavirus COVID-19 epidemic in South Korea based on recovered and death cases. *Journal of Medical Virology*, **92**, 1603–1608.

Anastassopoulou, C., Russo, L., Tsakris, A. and Siettos, C. (2020) Data-based analysis, modelling and forecasting of the COVID-19 outbreak. *PLOS ONE*, **15**, 1–21.

Bengtsson, L., Gaudart, J., Lu, X., Moore, S., Wetter, E., Sallah, K., Rebaudet, S. and Piarroux, R. (2015) Using mobile phone data to predict the spatial spread of cholera. *Scientific Reports*, **5**.

Chintalapudi, N., Battineni, G. and Amenta, F. (2020) COVID-19 virus outbreak forecasting of registered and recovered cases after sixty day lockdown in Italy: A data driven model approach. *Journal of Microbiology, Immunology and Infection*, **53**, 396–403.

Doshi, P. (2011) The elusive definition of pandemic influenza. *Bulletin of the World Health Organization*, **89**, 532–538.

Dudley, J. P. and Lee, N. T. (2020) Disparities in age-specific morbidity and mortality from SARS-CoV-2 in China and the Republic of Korea. *Clinical Infectious Diseases*, **71**, 863–865.

Grasselli, G., Pesenti, A. and Cecconi, M. (2020) Critical care utilization for the COVID-19 outbreak in Lombardy, Italy: Early experience and forecast during an emergency response. *JAMA*, **323**, 1545–1546.

Hallal, P. C., Hartwig, F. P., Horta, B. L., Silveira, M. F., Struchiner, C. J., Vidaletti, L. P., Neumann, N. A., Pellanda, L. C., Dellagostin, O. A., Burattini, M. N. et al. (2020) SARS-CoV-2 antibody prevalence in Brazil: Results from two successive nationwide serological household surveys. *The Lancet Global Health*, **8**, e1390–e1398.

Jin, J.-M., Bai, P., He, W., Wu, F., Liu, X.-F., Han, D.-M., Liu, S. and Yang, J.-K. (2020) Gender differences in patients with COVID-19: Focus on severity and mortality. *Frontiers in Public Health*, **8**, 152.

Kronbichler, A., Kresse, D., Yoon, S., Lee, K. H., Effenberger, M. and Shin, J. I. (2020) Asymptomatic patients as a source of COVID-19 infections: A systematic review and meta-analysis. *International Journal of Infectious Diseases*, **98**, 180–186. URLhttp://www.sciencedirect.com/science/article/pii/S1201971220304872.

Last, J. M. (2001) *A Dictionary of Epidemiology*. New York: Oxford University Press, 4th edn.

Laurens Holmes, J., Enwere, M., Williams, J., Ogundele, B., Chavan, P., Piccoli, T., Chinaka, C., Comeaux, C., Pelaez, L., Okundaye, O., Stalnaker, L., Kalle, F., Deepika, K., Philipcien, G., Poleon, M., Ogungbade, G., Elmi, H., John, V. and Dabney, K. W. (2020) Black–white risk differentials in COVID-19 (SARS-COV2) transmission, mortality and case fatality in the United States: Translational epidemiologic perspective and challenges. *Int. J. Environ. Res. Public Health*, **17**, 4322.

Manski, C. F. and Molinari, F. (2020) Estimating the COVID-19 infection rate: Anatomy of an inference problem. *Journal of Econometrics*.

Moghadas, S. M., Shoukat, A., Fitzpatrick, M. C., Wells, C. R., Sah, P., Pandey, A., Sachs, J. D., Wang, Z., Meyers, L. A., Singer, B. H. and Galvani, A. P. (2020) Projecting hospital utilization during the COVID-19 outbreaks in the United States. *Proceedings of the National Academy of Sciences*, **117**, 9122–9126.

Obukhov, A. G., Stevens, B. R., Prasad, R., Calzi, S. L., Boulton, M. E., Raizada, M. K., Oudit, G. Y. and Grant, M. B. (2020) SARS-CoV-2 infections and ACE2: Clinical outcomes linked with increased morbidity and mortality in individuals with diabetes. *Diabetes*, **69**, 1875–1886.

Oliver, N., Lepri, B., Sterly, H., Lambiotte, R., Deletaille, S., Nadai, M. D., Letouzé, E., Salah, A. A., Benjamins, R., Cattuto, C., Colizza, V., de Cordes, N., Fraiberger, S. P., Koebe, T., Lehmann, S., Murillo, J., Pentland, A., Pham, P. N., Pivetta, F., Saramäki, J., Scarpino, S. V., Tizzoni, M., Verhulst, S. and Vinck, P. (2020) Mobile phone data for informing public health actions across the COVID-19 pandemic life cycle. *Science Advances*, **6**.

Omori, R., Mizumoto, K. and Chowell, G. (2020) Changes in testing rates could mask the novel coronavirus disease (COVID-19) growth rate. *International Journal of Infectious Diseases*, **94**, 116–118.

Pollán, M., Pérez-Gómez, B., Pastor-Barriuso, R., Oteo, J., Hernán, M. A., Pérez-Olmeda, M., Sanmartín, J. L., Fernández-García, A., Cruz, I., de Larrea, N. F. et al. (2020) Prevalence of SARS-CoV-2 in Spain (ENE-COVID): A nationwide, population-based seroepidemiological study. *The Lancet*, **396**, 535–544.

Rainisch, G., Undurraga, E. A. and Chowell, G. (2020) A dynamic modeling tool for estimating healthcare demand from the COVID19 epidemic and evaluating population-wide interventions. *International Journal of Infectious Diseases*, **96**, 376–383.

Salje, H., Kiem, C. T., Lefrancq, N., Courtejoie, N., Bosetti, P., Paireau, J., Andronico, A., Hozé, N., Richet, J., Dubost, C.-L., Strat, Y. L., Lessler, J., Levy-Bruhl, D., Fontanet, A., Opatowski, L., Boelle, P.-Y. and Cauchemez, S. (2020) Estimating the burden of SARS-CoV-2 in France. *Science*, **369**, 208–211. URLhttps://science.sciencemag.org/content/369/6500/208.

Tizzoni, M., Bajardi, P., Decuyper, A., King, G. K. K., Schneider, C. M., Blondel, V., Smoreda, Z., González, M. C. and Colizza, V. (2014) On the use of human mobility proxies for modeling epidemics. *PLOS Computational Biology*, **10**, 1–15. URLhttps://doi.org/10.1371/journal.pcbi.1003716.

Tromberg, B. J., Schwetz, T. A., Pérez-Stable, E. J., Hodes, R. J., Woychik, R. P., Bright, R. A., Fleurence, R. L. and Collins, F. S. (2020) Rapid scaling up of Covid-19 diagnostic testing in the United States — the NIH RADx initiative. *New England Journal of Medicine*, **383**, 1071–1077.

Wesolowski, A., Qureshi, T., Boni, M. F., Sundsøy, P. R., Johansson, M. A., Rasheed, S. B., Engø-Monsen, K. and Buckee, C. O. (2015) Impact of human mobility on the emergence of dengue epidemics in Pakistan. *Proceedings of the National Academy of Sciences*, **112**, 11887–11892.

Yadav, M., Perumal, M. and Srinivas, M. (2020) Analysis on novel coronavirus (COVID-19) using machine learning methods. *Chaos, Solitons & Fractals*, **139**, 110050.

Part II

Modelling

3

Basic epidemiological features

Dani Gamerman

Universidade Federal de Minas Gerais/Universidade Federal do Rio de Janeiro, Brazil

Juliana Freitas

Universidade Federal de Minas Gerais, Brazil

Leonardo Nascimento

Universidade Federal do Amazonas, Brazil

CONTENTS

This chapter describes the main underlying features of epidemics and pandemics modelling. The epidemiological origins of these features are presented and discussed. Extensions and variations are described and compared. These give rise to a variety of modelling options. The properties that help characterise these models are also introduced. Finally, the adequacy of these choices in real pandemic data is illustrated and discussed.

3.1   Introduction and main ideas

Our main aim in this book is to deal with epidemic/pandemic data. A description of pandemic data was already presented in Chapter 2. Here, we reinforce that, as mentioned in Chowell et al. (2019), the dynamic of the spread of cases depends on several factors, such as the form of infection, reproducibility power, cultural/social habits, among others. Nevertheless, although the

DOI: 10.1201/9781003148883-3

course of a pandemic is varied, an overall pattern of the temporal evolution is expected.

The idea of the overall pattern is that the counts of cases will initially increase with time, as more and more people are exposed to the infected cases. In other words, we expect an increasing behaviour at the beginning of the epidemic. As more people become infected, there are fewer and fewer people that these individuals can infect. Thus, counts of new cases will begin to decelerate until a moment at which the maximum number of cases is observed. This moment may be referred to as the pandemic peak. As usually happens, a decrease in the number of cases should steadily take place (Hsieh and Ma, 2009). It is worth mentioning that there can be many variations in the dynamics of this evolution. For example, there can be a change in the definition of a case, social distancing measures usually imposed on infected populations may change, new surges of contagion may take place, just to cite a few. These issues are handled in the coming chapters and are also discussed in Chowell et al. (2019), and in Roosa et al. (2020).

The described pandemic path can be accommodated in functional forms over time indexed by a vector of parameters. These functions are referred to as growth curves (Pearl and Reed, 1925; von Bertalanffy, 1957; Richards, 1959; Birch, 1999). These curves have been used in several applications, such as the height of plants (Richards, 1959; Meng et al., 1997; Birch, 1999), weight of animals (Pearl and Reed, 1925), as well as to model epidemic/pandemic data of diseases like Dengue (Hsieh and Ma, 2009; Hsieh and Chen, 2009), Ebola (Pell et al., 2018; Chowell et al., 2019), H1N1 (Hsieh et al., 2010), Plague (Chowell et al., 2019), SARS (Hsieh and Cheng, 2006; Chowell et al., 2019), and, more recently, COVID-19 (Roosa et al., 2020; Schumacher et al., 2021).

Evidently, there are many other ways of modelling the evolution pattern of cases in epidemics. Alternatives include compartmental models and their extensions. As already anticipated in Section 1.1.1, this book is devoted to the data-driven modelling approach to epidemics and thus this route will not be pursued here. The reader interested in compartmental models should refer to Kermack and McKendrick (1927), Wang et al. (2012), Chowell et al. (2016a), Chowell et al. (2016b), Yamana et al. (2016), Held et al. (2019), and Anastassopoulou et al. (2020), to cite just a few examples of a very large body of literature.

Now, turning the attention back to growth models, perhaps the simplest growth curve comes from the exponential growth model. According to Cramer (2002), this model has its origin in modelling population growth and it is defined via the differential equation

$$\frac{dM(t)}{dt} = \alpha M(t), \text{ subject to } M(0) = L_I, \tag{3.1}$$

where $M(t)$ represents the cumulative number of cases until time t, and α and L_I are positive constants. The idea behind (3.1) is that the rate of growth

$dM(t)/dt$ of the number of cases is proportional to the number $M(t)$ of infected individuals. The larger the infected population is, the more new infected cases will occur. The solution of this differential equation is given by

$$M(t) = a \exp(ct), \qquad (3.2)$$

where $a = L_I$ and $c = \alpha$. More about differential equations can be found in Boyce et al. (2017) and the derivation of this result can be seen in this chapter's appendix. Thus, under the exponential growth model, the number of cases grows exponentially with a rate c. In addition, we can note that (i) the initial condition is such that $M(0) = a$, and (ii) when $t \longrightarrow \infty$, $M(t) \longrightarrow \infty$, i.e., as the time evolves, the cumulative number of cases grows indefinitely. The last characteristic above may be unrealistic in many scenarios, since susceptible populations are usually finite (von Bertalanffy, 1957; Cramer, 2002). Moreover, this function presents no critical point; thus, it fails to represent the anticipated deceleration on the infection pattern.

Several growth curves have been proposed to provide flexibility and to accommodate features such as asymmetry, exponential growth at lower cases, finite limits, multiple peaks, and seasonality (Birch, 1999). The most immediate of these proposals is the logistic growth model (Verhulst, 1838, 1844, 1847). The differential equation that originates this model is given by

$$\frac{dM(t)}{dt} = \alpha M(t) \left(1 - \frac{M(t)}{\beta}\right), \text{ subject to } M(0) = L_I, \qquad (3.3)$$

where α, β and L_I are positive constants. The intuition behind model (3.3) is the assumption that the rate of growth of cases is proportional to (i) the number $M(t)$ of cases that occur until time t, and (ii) the proportion of population individuals that are still exposed to infection (given by $1 - M(t)/\beta$) (Cramer, 2002). Thus, β represents the total number that will eventually get infected, i.e., $M(\infty) = \beta$. The solution to (3.3) is given by

$$M(t) = \frac{a \exp\{ct\}}{1 + b \exp\{ct\}}, \qquad (3.4)$$

where $a = \beta L_I/(\beta - L_I)$, $b = L_I/(\beta - L_I)$, and $c = \alpha$. More details about these results and the relationship between the vector (a, b, c) and the vector (α, β, L_I) can be found in this chapter's appendix. Still, it is clear that a and b are related to each other, since both of them are functions of β and L_I. It is worth pointing out that when $b = 0$, the exponential growth model is obtained as a special case. Also, $M(0) = L_I = a/(1 + b)$.

Figure 3.1 provides an illustration of the logistic growth curve. The figure draws attention to the fact that the logarithmic representation is useful to

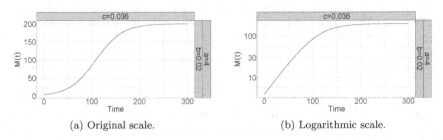

(a) Original scale.   (b) Logarithmic scale.

FIGURE 3.1: Illustration of the logistic growth curve for the cumulative cases until time t when $a = 4$, $b = 0.02$, and $c = 0.036$, both on the original (Panel 3.1a) and the logarithmic scales (Panel 3.1b).

investigate possible log-linearity of parts of the curve, especially in the beginning. In addition to that, it helps with interpretation when large variations in the original scale occur, which is frequently the case.

Then, observing Panel 3.1b we can see that, in this hypothetical case, the epidemic is starting at time 0 with $M(0) = a/(1 + b) = 3.922 \approx 4$ confirmed cases. Moreover, the curve stabilises at a total of around 200 cases.

Another interesting point about growth curves is that it is possible to interpret the pattern of the new arriving cases. Analysis of growth curves in this form may deliver a clearer view of the course of the epidemic. In this case we must simply take the derivative of the function $M(t)$, which is denoted by $\mu(t)$. Considering the logistic growth model, $\mu(t)$ is given by

$$\mu(t) \quad = \quad \frac{ac\exp\{ct\}}{[1 + b\exp\{ct\}]^2}. \tag{3.5}$$

The scenario of Figure 3.1 is revisited in Figure 3.2, in the form of the new cases at each time t.

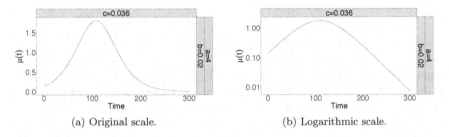

(a) Original scale.   (b) Logarithmic scale.

FIGURE 3.2: Illustration of the logistic growth curve for the new cases at time t when $a = 4$, $b = 0.02$, and $c = 0.036$, both on the original (Panel 3.2a) and the logarithmic scales (Panel 3.2b).

Figure 3.2 allows discussion about the course of an epidemic more clearly. A potential drawback involving the logistic curve is the imposition of symmetry with respect to the peak of new cases. The symmetry in this case means that the number of cases that were observed l days before the peak is exactly the same as the number of cases observed l days after the peak date, for any l. This characteristic can be observed in Figures 3.1a and especially in 3.2a. As a consequence, the total number of cases before and after the peak must be equal. Such feature may be restrictive and it also may seem an unrealistic situation in many cases, as mentioned in Pearl and Reed (1925), Birch (1999), and Tsallis and Tirnakli (2020).

Different strategies to accommodate asymmetric patterns in growth curves have been proposed in the literature. For example, Birch (1999) included both quadratic and cubic terms of time. In a somewhat similar strategy, Richards (1959) and Tsallis and Tirnakli (2020) used additional forms to model the dependence on time of the temporal evolution of cases. These and other strategies to extend the logistic growth model are presented in the next section.

3.2 Model extensions

Having the logistic growth model as a starting point, several authors proposed extensions to this model. One of the extensions that are most commonly used is the Richards growth model, often also called the generalised logistic model. Richards proposed a generalisation of the logistic model, given by

$$\frac{dM(t)}{dt} = \alpha M(t) \left[1 - \left(\frac{M(t)}{\beta} \right)^{\gamma} \right], \text{ subject to } M(0) = L_I, \qquad (3.6)$$

where α, β, γ and L_I are positive constants. It is interesting to note that (3.6) can be re-written as $\frac{dM(t)}{dt} = \frac{\alpha}{\beta^{\gamma}} M(t) \left(\beta^{\gamma} - (M(t))^{\gamma} \right)$, leading to an interpretation qualitatively similar to (3.3). Then, solving Equation (3.6), leads to

$$M(t) = \frac{a}{[b + \exp(-ct)]^f}, \qquad (3.7)$$

where we defined $a = (\beta L_I)/(\beta^{\gamma} - L_I^{\gamma})^{1/\gamma}$, $b = (L_I^{\gamma})/(\beta^{\gamma} - L_I^{\gamma})$, $c = \alpha \gamma$, and $f = 1/\gamma$. As a result, the relation between the components of the parameter vector (a, b, c, f) can be seen. Here, all the components of this vector depend on γ. Yet, a and b also depend on β and on L_I.

In this growth function, the parameter f is responsible for incorporating the asymmetry. When $f = 1$, the Richards curve equals the logistic curve (see

Equation (3.4)). Then, as can be expected for this case, the vector of parameters (a, b, c) are also equal between these models. This equivalence also means that when $f = 1$, the Richards curve presents a symmetric shape in relation to the peak date. This growth curve has several other curves as special cases, like the autocatalytic, exponential, Gompertz, and monomolecular growth curves (Richards, 1959; Henderson et al., 2006; Panik, 2014). Another remark about this model is that, under the new parametrisation, the initial condition can be rewritten as $M(0) = a/(1 + b)^f$. For more details about the calculations involving this model, the reader is encouraged to see this chapter's appendix.

Figure 3.3 allows a visualisation of the possible forms that the Richards curve can assume. Note that the time axis is the same in all sub-figures. On the other hand, the variation of the number of cases between the sub-figures is quite large. Thus, the Richards growth model is able to contemplate situations that go from small epidemics—with a few reported cases—to worldwide epidemics like the COVID-19.

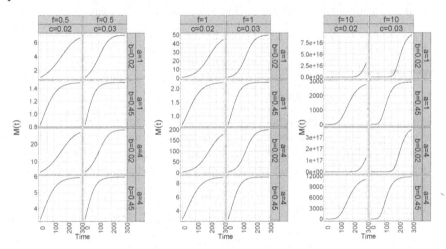

FIGURE 3.3: Illustration of the several shapes that the generalised logistic curve for the cumulative cases can assume. From the left to the right, the parameters c and f vary in the sets $\{0.02, 0.03\}$ and $\{0.5, 1, 10\}$, respectively. The curves in the middle column represent the logistic curve ($f = 1$). From the top to the bottom lines, the parameter a varies in the set $\{1, 4\}$ and the parameter b in the set $\{0.02, 0.45\}$.

When we are interested in analysing new cases, we find that

$$\mu(t) = M'(t) = \frac{acf \exp(-ct)}{[b + \exp(-ct)]^{(f+1)}}, \tag{3.8}$$

where $\mu(t)$ is the function that represents the counts of new cases at time t. Figure 3.4 reproduces the growth curves for the new counts based on the same growth curves that were obtained in Figure 3.3 for the cumulative counts.

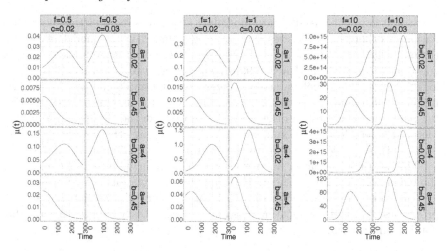

FIGURE 3.4: Illustration of the several shapes that the generalised logistic curve for the new cases can assume. From the left to the right, the parameters c and f vary in the sets $\{0.02, 0.03\}$ and $\{0.5, 1, 10\}$, respectively. The curves in the middle column represent the logistic curve ($f = 1$). From the top to the bottom lines, the parameter a varies in the set $\{1, 4\}$ and the parameter b in the set $\{0.02, 0.45\}$.

The Richards growth model has been extensively used in growth models, either as the main option or as a competitive alternative (Meng et al., 1997; Birch, 1999; Chowell et al., 2019, 2020). The reason for its popularity is the flexibility and its interesting properties. Some of these properties were previously mentioned, like asymmetry and the possibility of contemplating exponential growth at a lower number of cases. It also has a list of growth models as special cases. However, some authors like Birch (1999) and Wang et al. (2012) claim that Richards curve presents issues, such as lack of identifiability and computational instability. Some of these issues may be related to the estimation procedure and to the reparametrisation of the curve. Moreover, the parameter controlling the skewness does not present a biological meaning. The interpretation may be important for some applications.

Another interesting growth curve is the Janoschek model. This model can be compared to the Richards curve as both of these models represent generalisations of the logistic curve. Henderson et al. (2006) argue that the Janoschek model is similar to the Richards curve with respect to flexibility, but it has an advantage of computational stability. The equation for the cumulative cases for this model is

$$M(t) = a(1 - b \exp\{-ct^f\}), \text{ for } f > 1 \tag{3.9}$$

where a, b, and c are positive constants and $f > 1$.

Still, it is possible that a data modelling processing requires the accommodation of other characteristics that are not included in the models we mentioned earlier. For this and other reasons, alternative growth models have emerged. More recently, Birch (1999) proposed another generalisation of the logistic curve. Nevertheless, the outcomes using Richards growth model have better performance in some of cases studied in the mentioned paper. For the reader with interest in other growth curves—like the Gompertz, Morgan-Mercer-Floden, Negative Exponential, Weibull, among others—and the properties of these models, we refer to Henderson et al. (2006) and Panik (2014).

This section presented an overview of the main growth functions, and the equations that define them. Thus, all these points may be considered in the choice of the growth function to be used. The next section focuses on properties of epidemiological models and further interpretations that can be obtained from them.

3.3   Properties of epidemiological models

In the previous section, examples of growth curves were introduced and discussed. In addition to that, we described their importance, their origin and the meaning of the parameters that index these models. Then, proceeding deeper with the details of these models, once the values of the parameters of a growth model are known, several useful tools to characterise them can be derived. Some of these characteristics are the reproduction number, the peak date, the number of cases at the peak, and the total number of cases at the end of the epidemic/pandemic. In this section, all these characteristics will be defined, along with an indication of their importance. These will be illustrated for some of the growth curves of the previous section, to highlight the changes that may take place from epidemic to epidemic.

As we have already anticipated, the modelling process with growth curves allows us to consider data in the form of the cumulative cases at each time, or the daily (weekly, monthly, etc.) new cases. Then, the first property refers to the relationship between the growth curves for each of these two types of model specification, already presented in Section 3.1.

Definition 1 (Growth curve for new cases) *The growth curve for the new cases in time t is given by the derivative of the growth curve for the cumulative number of cases $M(t)$. We represent this curve by $\mu(t) = M'(t)$.*

In a strict sense, the cumulative counts of cases until time t are composed of the sum of the new cases in each time point. Therefore, since the curve $M(t)$ represent these counts, it would be more appropriate to set the curve for the new cases as the difference $M(t) - M(t-1)$. Alternatively, one could use an approximation for this difference, given by the rate of growth $\mu(t) =$

$M'(t)$ described in Definition 1, which is more computationally amenable. In this case, the cumulative counts could be accordingly modified to $M(t) = \sum_{j=1}^{t} \mu(j)$ for consistency.

The function $\mu(t)$ for the logistic and the generalised logistic growth models are given by Equations (3.5) and (3.8), respectively. We recall that an evaluation of the growth curve in this form is provided in Figure 3.4. This representation is also useful to define other properties like the epidemic peak date (see Definition 2 below).

The second definition concerns the epidemic peak. The peak date will take place in a time t such that the maximum number of new cases has been reached. After such date, the number of new cases will decrease until the epidemic ends.

Definition 2 (Epidemic peak) *The peak date indicated by an epidemic growth model, denoted by t_{peak}, is given by the time t such that $\mu'(t) = 0$.*

There is a closed form expression for this date for the generalised logistic model. It is provided by noting that

$$\mu'(t) = \frac{\exp(-ct)[-b + f \exp(-ct)]}{[b + \exp(-ct)]^{(f+2)}} = 0 \text{ implies that } t_{peak} = \frac{1}{c} \log\left(\frac{f}{b}\right).$$

$$(3.10)$$

Consequently, $t_{peak} = (1/c) \log(1/b)$ for the logistic curve.

In what follows, Definition 3 refers to the total cases that are expected to happen by the end of the epidemic curve.

Definition 3 (Total number of cases) *The total number of cases is what is expected to be accumulated in cases by the end of the epidemic curve. In mathematical terms, this number is obtained by taking the limit of $M(t)$ when $t \longrightarrow \infty$.*

For the generalised logistic, it is easy to obtain that

$$\lim_{t \longrightarrow \infty} \frac{a}{[b + \exp(-ct)]^f} = \frac{a}{b^f}.$$

$$(3.11)$$

When $f = 1$, in the logistic model, this quantity turns out to be a/b. Note that the parameter c has no direct influence on the total number of cases by the end of the epidemic. It only influences the amount of time it takes to get there. Also, it is worth recalling that for the exponential growth model, the number of cases grows indefinitely. Therefore, we are not able to obtain the peak date nor the total number of cases.

Considering Definition 3, the end of a pandemic will take place only as $t \longrightarrow \infty$. However, for practical purposes, there may be an interest in having a reference finite date to characterise closeness to the end. One possibility is to consider the time t at which a sufficiently large proportion p of the total cases occurred. Typical values for p range from 95% to 99.9%, but the choice is case specific and may also depend on the use that will be made of this quantity. This practical version of the total number of cases is frequently used as a proxy to inform the approach of the end of a pandemic.

Definition 4 (End date) *The end date of an epidemic, denoted by $t_{end,100p\%}$, is defined as the time at which a sufficiently large proportion p of the total number of cases has already occurred.*

Additional features concern quantities related to the peak date. For example, once we have the peak date, we may be interested in knowing the number of cases and also the cumulative number of cases at this very date. For the generalised logistic model, these quantities are given by

$$\mu(t_{peak}) = \frac{acf \exp(-ct_{peak})}{[b + \exp(-ct_{peak})]^{(f+1)}} = \frac{a}{b^f} \frac{c}{\left(1 + \dfrac{1}{f}\right)^{(f+1)}}$$

and

$$M(t_{peak}) = \frac{a}{[b + \exp(-ct_{peak})]^f} = \frac{a}{b^f} \frac{1}{\left(1 + \dfrac{1}{f}\right)^f}.$$

Note that the quantities above are a fraction of the total number of cases in Definition (3). In order to obtain the expressions for the logistic curve, we must simply take $f = 1$ in the equations above. Then, if $f = 1$, $M(t_{peak}) = 0.5a/b$ is half of the total number of cases, a direct consequence of the symmetry of the logistic curve.

Table 3.1 displays the values of all the quantities described above for the generalised logistic growth models displayed in Figures 3.3 and 3.4.

This table summarises some of the discussion of this section. Even though none of the combinations are the same, some equal representations of the peak date are obtained, for example. This similarity happened with other quantities displayed in this table; it is obtained because the parameters have similar roles in some—but not all—of the definitions above. Finally, the last examples exhibit very large numbers of total cases. This shows the ability of this curve to model the pandemics in general.

The last definition to be presented is the reproduction number. Heesterbeek and Dietz (1996), Heffernan et al. (2005), and Hsieh and Ma (2009) highlight the importance of this quantity, mainly in epidemiological studies. The reproduction number varies in time and informs us about the spread of

TABLE 3.1: Information about the epidemiological characteristics of the growth models illustrated in Figures 3.3 and 3.4. (Table entries, except for the vector of parameters, are rounded to integers.)

Parameters				Peak			End		
a	b	c	f	t_{peak}	$\mu(t_{peak})$	$M(t_{peak})$	$t_{end,95\%}$	$t_{end,99.9\%}$	$M(\infty)$
1	0.02	0.02	0.5	161	0	4	298	> 300	7
1	0.45	0.02	0.5	5	0	1	21	30	1
4	0.02	0.02	0.5	161	0	16	298	> 300	28
4	0.45	0.02	0.5	5	0	3	158	> 300	6
1	0.02	0.03	0.5	107	0	4	199	257	7
1	0.45	0.03	0.5	4	0	1	14	20	1
4	0.02	0.03	0.5	107	0	16	199	257	28
4	0.45	0.03	0.5	4	0	3	106	> 300	6
1	0.02	0.02	1	196	0	25	> 300	> 300	50
1	0.45	0.02	1	40	0	1	129	150	2
4	0.02	0.02	1	196	1	100	> 300	> 300	200
4	0.45	0.02	1	40	0	4	202	> 300	9
1	0.02	0.03	1	130	0	25	229	> 300	50
1	0.45	0.03	1	27	0	1	86	100	2
4	0.02	0.03	1	130	2	100	229	> 300	200
4	0.45	0.03	1	27	0	4	135	> 300	9
1	0.02	0.02	10	311	6.85×10^{14}	3.77×10^{16}	> 300	> 300	9.77×10^{16}
1	0.45	0.02	10	155	21	1132	> 300	> 300	2937
4	0.02	0.02	10	311	2.74×10^{15}	1.51×10^{17}	> 300	> 300	3.91×10^{17}
4	0.45	0.02	10	155	82	4529	> 300	> 300	11747
1	0.02	0.03	10	207	1.03×10^{15}	3.77×10^{16}	> 300	> 300	9.77×10^{16}
1	0.45	0.03	10	103	31	1132	203	> 300	2937
4	0.02	0.03	10	207	4.11×10^{15}	1.51×10^{17}	> 300	> 300	3.91×10^{17}
4	0.45	0.03	10	103	124	4529	203	> 300	11747

the disease in terms of infected cases, which, in turn, gives an indication of the epidemic evolution. The origin of this quantity, however, is not related to epidemics; one can find more about the history of this number in Heesterbeek and Dietz (1996). The definition of the reproduction number is given below.

Definition 5 (Reproduction number $R(t)$) *The reproduction number is defined as the average number of new infections that any infected individual produces. This idea can be represented as*

$$R(t) = \frac{\mu(t)}{M(t-1) - Q(t-1)}, \, t > 0, \qquad (3.12)$$

where $\mu(t)$ represents the curve for the number of new cases in time t, $M(t-1)$

is the cumulative number of cases at time $t - 1$, and $Q(t - 1)$ stands for the cumulative counts of removed (i.e., recovered and deceased) individuals at time $t - 1$.

The reproduction number has a reference mark of $R(t) = 1$. Heffernan et al. (2005) argues that this mark helps the understanding of the contrasting situations between an increasing number of infections *versus* a deceleration in the epidemic growth. Nonetheless, although $R(t)$ presents a consistent theoretical definition (Definition 5), there are different ways of putting the definition in mathematical terms and, therefore, into practice. According to the same authors, the form of writing may vary depending on the source and type of data as well as the modelling process being applied. The types of data may be cell reproduction, population growth, as well as disease dynamics, which is our target. Also, it may depend on how much data-specific information—such as infection duration, possibility of reinfection, fatality ratio—is available. Heesterbeek and Dietz (1996) and Heffernan et al. (2005) presented an interesting discussion about this number, how to interpret it, and different forms of translating $R(t)$ in mathematical equations. In the latter referred work, there is a brief section focusing on the cases where the model is based on growth curves. In addition to that, both works of Hsieh and Ma (2009) and Hsieh et al. (2010) had the same strategy to obtain $R(t)$ at initial times using the generalised logistic model, for example. However, it depended on the infection period of the disease. In the case of an outbreak of a novel disease, this information may not be known. So, in Equation (3.12) we present another possible way to estimate $R(t)$ using growth curves.

Figure 3.5 shows examples of the evolution in time of the reproduction number $R(t)$ based on the logistic and the generalised logistic growth models. The values of the vector of parameters of the models that compose this figure are the same as Figures 3.3 and 3.4. We highlight the reference mark of $R(t) = 1$, $t > 0$ represented by the dotted line.

An overall interpretation of the curves in Figure 3.5 is the evaluation regarding the reference value of $R(t) = 1$. Since in the beginning of an epidemic the number of cases will probably have an increasing behaviour, we can expect $R(t)$ to be greater than 1. This pattern remains unchanged until the number of new cases decelerates, by which time window $R(t)$ will eventually fall below 1. This pattern will only be reinforced after passing the peak date. These details can be checked by referring back to Figure 3.4. In addition to that, we note that the largest values that $R(t)$ assume are on the third block of subfigures. This block is exactly the situation when a very large number of infected people by the end of the epidemic is observed. This conclusion is clearer in Figure 3.3 and in Table 3.1.

In this section we focused on the main properties of epidemiological growth models. Other interesting and useful properties can be found in Richards (1959). All the calculations described above can be performed over other growth curves like the von Bertalanffy (1957), and Janoschek, for example.

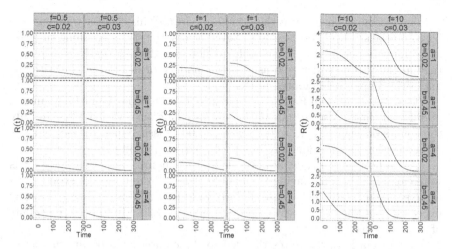

FIGURE 3.5: Illustration of the several shapes that the reproduction number $R(t)$ can assume based on the generalised logistic curve. This representation is based on Equation (3.12) with $Q(t-1) = 0.9M(t-1)$. From the left to the right, the parameters c and f vary in the sets $\{0.02, 0.03\}$ and $\{0.5, 1, 10\}$, respectively. The curves in the middle column represent the logistic curve ($f = 1$). From the top to the bottom lines, the parameter a varies in the set $\{1, 4\}$ and the parameter b in the set $\{0.02, 0.45\}$. The dotted lines represent the reference mark of $R(t) = 1$.

The next section addresses how appropriate these models are in the context of pandemic data.

3.4 Are these models appropriate?

The models seem reasonable from the justification for their use, but do they perform well? At this preliminary level one could be satisfied with the knowledge that the models are minimally capable of picking the temporal signal, the overall trend presented by the data, after appropriately discarding observation noise. There are many ways to define model performance. Here we will rely on visual inspection as a first, basic reality check. Some more sound performance criteria are discussed and applied in Part III of this book.

For this evaluation, we considered COVID-19 data from Belgium, Canada, Italy, Japan, and South Africa. This choice was made aiming at providing an overview of different continents of the world. Confirmed COVID-19 cases were considered and the data was collected from the data-base prepared by the Center for Systems Science and Engineering at Johns Hopkins University.

The reasoning that led to the growth models of this chapter were entirely based on the study of disease cases. Nevertheless, the same reasoning suggests these models might also be suitable to describe disease deaths. Therefore, they might also be considered as possible models for deaths, in line with the data-driven approach adopted in this book.

It can be noted from Figure 3.6 that the COVID-19 pandemic hit countries at different times. Also the magnitude of this pandemic is varied: while in Italy the counts of new cases in a day reached 6,000 cases, Japan did not surpass 700 confirmed cases in any day in the comparison period.

The important message, however, is that the figure shows that the models were capable of accompanying the main features of the temporal evolution of these pandemic data. This is particularly valid for the trend of disease cases as expected. But they also seem to provide adequate description for the counts of disease deaths.

The estimation procedures that led to these curves will be the subject of Chapter 6. In addition, comparing the two curve fittings, we can note the importance of accommodating the asymmetry, essentially for Belgium, Canada, and Italy. The peaks seem to have been better identified when employing the generalised logistic model. In the case of countries where counts seem to exhibit a symmetric behaviour, both fitted curves are similar, as expected. This was the case of Japan and South Africa.

This section suggested that the (generalised) logistic model might be a suitable platform to consider for modelling the signal from pandemic data, after observational noise was cleared. This was based on the evidence reported in Figure 3.6 but experiments with other regions (not reported here) point in the same direction.

However, there will still be the need to include other specific characteristics in the modelling process. Some of these additional data characteristics will be presented and discussed in Chapters 4 and 5. The following appendix details the main calculations presented in this chapter. The next chapter focuses on explaining the distributions that can be associated with pandemic data and how they can be related to the growth curves described here.

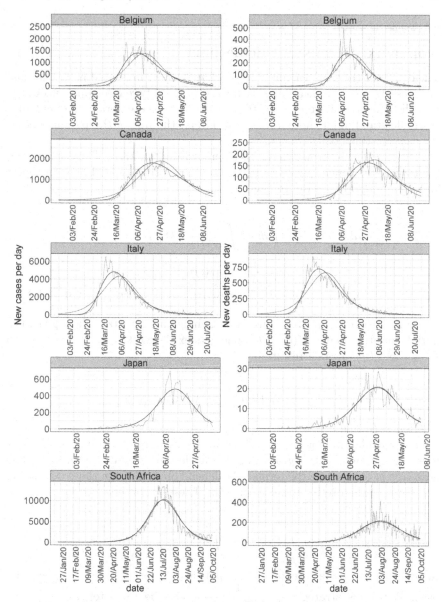

FIGURE 3.6: Results from the fitted means of the logistic (black dotted line) and the generalised logistic (black straight line) models applied to COVID-19 data (straight grey line) for Belgium, Canada, Italy, Japan, and South Africa. Left panel contains the confirmed cases and the right panel shows confirmed deaths. Data source: data-base prepared by the Center for Systems Science and Engineering at Johns Hopkins University.

Appendix: Details of calculations

The details of the main calculations of Chapter 3 are presented here. Our goal in this section is to indicate the path to solve the differential equations of the growth models that were presented. In addition, these details may enhance the understanding of the parametrisation and the role of the parameters we are using in this book.

The exponential growth model is given by Equation (3.1). Then, separating the terms that depend on $M(t)$ from those terms that depend only on time t and then integrating on both sides, gives

$$dt = \frac{1}{\alpha M(t)} dM(t) \quad \Rightarrow \quad \int dt = \int \frac{1}{\alpha M(t)} dM(t)$$

$$\Rightarrow \quad t = \frac{1}{\alpha} \log\left(M(t)\right) + K$$

$$\Rightarrow \quad M(t) = \exp\{\alpha(t - K)\} = \exp\{\alpha t\}\exp\{-\alpha K\},$$

where K is the constant resulting from an indefinite integral. Then, given the initial condition being $M(0) = L_I$, one gets $\exp\{-\alpha K\} = L_I$, which leads to $K = -\log\left(L_I\right)/\alpha$. As a result,

$$M(t) = \exp\{\alpha t\}\exp\{-\alpha K\} = a \exp\{ct\},$$

where $a = L_I$ and $c = \alpha$.

The calculations for the logistic growth curve are quite similar. The differential equation for this growth curve was described by Equation (3.3). The next step is to separate the terms that depend on $M(t)$ from those depending only on time, and to integrate both sides. These operations lead to

$$dt = \frac{\beta}{\alpha M(t)(\beta - M(t))} dM(t) \quad \Rightarrow \quad \int dt = \frac{\beta}{\alpha} \int \frac{1}{M(t)(\beta - M(t))} dM(t).$$

In order to solve the integral above, partial fraction decomposition can be used. With this strategy, $\dfrac{1}{M(t)(\beta - M(t))}$ can be rewritten as $\dfrac{1}{\beta M(t)} + \dfrac{1}{\beta(\beta - M(t))}$. With this modification, the integral can be easily solved as

$$t = \frac{1}{\alpha} \ln\left(\frac{M(t)}{\beta - M(t)}\right) + K \Rightarrow M(t) = \frac{\beta \exp\{\alpha t\}\exp\{-\alpha K)\}}{[1 + \exp\{\alpha t\}\exp\{-\alpha K)\}]},$$

where K is the constant from the indefinite integral. Given the initial condition $M(0) = L_I$, the constant K can be found after solving $M(0) =$

$$\frac{\beta \exp\{-\alpha K\}}{[1 + \exp\{-\alpha K\}]} = L_I \text{ which implies } K = \frac{1}{\alpha} \ln\left(\frac{\beta - L_I}{L_I}\right).$$ Finally, defin-

ing $a = \beta \exp\{-\alpha K\} = \dfrac{\beta L_I}{\beta - L_I}$, $b = \exp\{-\alpha K\} = \dfrac{L_I}{\beta - L_I}$, and $c = \alpha$ gives Equation (3.4).

The third growth curve we presented in this chapter is the Richards, or generalised logistic, curve. This model was introduced by Equation (3.6). Repeating the ideas described above leads to

$$\int dt = \frac{\beta^\gamma}{\alpha} \int \frac{1}{M(t)(\beta^\gamma - (M(t))^\gamma)} dM(t). \tag{3.13}$$

One way of solving this integral is to apply the substitution technique with $u = (M(t))^\gamma$. This substitution leads the right-hand side of (3.13) to be $\dfrac{\beta^\gamma}{\alpha} \int \dfrac{1}{\gamma u(\beta^\gamma - u)} du$. After using the partial fraction decomposition, the integrand becomes $\dfrac{1}{\gamma u} + \dfrac{1}{\gamma(\beta^\gamma - u)}$. Finally, after following the steps of the previous calculation, the solution for Equation (3.13) is

$$t = \frac{1}{\alpha\gamma} \ln\left(\frac{(M(t))^\gamma}{\beta^\gamma - (M(t))^\gamma}\right) + K \text{ leading to}$$

$$M(t) = \frac{\beta \exp\{-\alpha K\}}{[\exp\{-\alpha\gamma K\} + \exp\{-\alpha\gamma t\}]^{1/\gamma}},$$

where K is, again, the constant resulting from an indefinite integral. Then, using the initial condition as $M(0) = L_I$ gives $M(0) = \dfrac{\beta \exp\{-\alpha K\}}{[1 + \exp\{-\alpha\gamma K\}]^{1/\gamma}} = L_I$ which implies that $K = -\dfrac{1}{\alpha\gamma} \ln\left(\dfrac{L_I^\gamma}{\beta^\gamma - L_I^\gamma}\right)$. Finally, the vector parameters used in Equation (3.7) are given by $a = \beta \exp\{-\alpha K\} = (\beta L_I)/(\beta^\gamma - L_I^\gamma)^{1/\gamma}$, $b = \exp\{-\alpha\gamma K\} = (L_I^\gamma)/(\beta^\gamma - L_I^\gamma)$, $c = \alpha\gamma$, and $f = 1/\gamma$.

Bibliography

Anastassopoulou, C., Russo, L., Tsakris, A. and Siettos, C. (2020) Data-based analysis, modelling and forecasting of the COVID-19 outbreak. *PLOS ONE*, **15**, 1–21.

von Bertalanffy, L. (1957) Quantitative laws in metabolism and growth. *The Quarterly Review of Biology*, **32**, 217–231.

Birch, C. P. D. (1999) A new generalized logistic sigmoid growth equation compared with the Richards growth equation. *Annals of Botany*, **83**, 713–723.

Boyce, W. E., DiPrima, R. C. and Meade, D. B. (2017) *Elementary Differential Equations*. John Wiley & Sons, 11 edn.

Chowell, G., Luo, R., Sun, K., Roosa, K., Tariq, A. and Viboud, C. (2020) Real-time forecasting of epidemic trajectories using computational dynamic ensembles. *Epidemics*, **30**, 100379.

Chowell, G., Sattenspiel, L., Bansal, S. and Viboud, C. (2016a) Mathematical models to characterize early epidemic growth: A review. *Physics of Life Reviews*, **18**, 66–97.

Chowell, G., Tariq, A. and Hyman, J. M. (2019) A novel sub-epidemic modeling framework for short-term forecasting epidemic waves. *BMC Medicine*, **17**.

Chowell, G., Viboud, C., Simonsen, L. and Moghadas, S. M. (2016b) Characterizing the reproduction number of epidemics with early subexponential growth dynamics. *Journal of The Royal Society Interface*, **13**, 20160659.

Cramer, J. (2002) The origins of logistic regression. *Tinbergen Institute Working Paper*, **2002-119/4**.

Heesterbeek, J. A. P. and Dietz, K. (1996) The concept of R_0 in epidemic theory. *Statistica Neerlandica*, **50**, 89–110.

Heffernan, J. M., Smith, R. J. and Wahl, L. M. (2005) Perspectives on the basic reproductive ratio. *J. R. Soc. Interface*, **2**, 281–293.

Held, L., Hens, N., O'Neill, P. and Wallinga, J. (eds.) (2019) *Handbook of Infectious Disease Data Analysis*. Boca Raton: Chapman and Hall/CRC, 1st edn.

Henderson, P., Seaby, R. and Somes, R. (2006) *Growth II*. 2006 Pisces Conservation Ltd, Lymington, England. URLwww.pisces-conservation.com.

Hsieh, Y.-H. and Chen, C. W. S. (2009) Turning points, reproduction number, and impact of climatological events for multi-wave dengue outbreaks. *Tropical Medicine & International Health*, **14**, 628–638.

Hsieh, Y.-H. and Cheng, Y.-S. (2006) Real-time forecast of multiphase outbreak. *Emerging Infectious Diseases*, **12**, 122–127.

Hsieh, Y.-H., Fisman, D. N. and Wu, J. (2010) On epidemic modeling in real time: An application to the 2009 novel A (H1N1) influenza outbreak in Canada. *BMC Res Notes*, **3**.

Hsieh, Y.-H. and Ma, S. (2009) Intervention measures, turning point, and reproduction number for dengue, Singapore, 2005. *The American Journal of Tropical Medicine and Hygiene*, **80**, 66–71.

Kermack, W. O. and McKendrick, A. G. (1927) A contribution to the mathematical theory of epidemics. *Proceedings of the Royal Society A*, **115**, 700–721.

Meng, F.-R., Meng, C. H., Tang, S. and Arp, P. A. (1997) A new height growth model for dominant and codominant trees. *Forest Science*, **43**, 348–354.

Panik, M. J. (2014) *Growth Curve Modeling: Theory and Applications*. Hoboken, New Jersey: John Wiley & Sons.

Pearl, R. and Reed, L. J. (1925) Skew-growth curves. *Proceedings of the National Academy of Sciences*, **11**, 16–22.

Pell, B., Kuang, Y., Viboud, C. and Chowell, G. (2018) Using phenomenological models for forecasting the 2015 Ebola challenge. *Epidemics*, **22**, 62–70. The RAPIDD Ebola Forecasting Challenge.

Richards, F. J. (1959) A flexible growth function for empirical use. *Journal of Experimental Botany*, **10**, 290–300.

Roosa, K., Lee, Y., Luo, R., Kirpich, A., Rothenberg, R., Hyman, J., Yan, P. and Chowell, G. (2020) Real-time forecasts of the COVID-19 epidemic in China from February 5th to February 24th, 2020. *Infectious Disease Modelling*, **5**, 256–263.

Schumacher, F. L., Ferreira, C. S., Prates, M. O., Lachos, A. and Lachos, V. H. (2021) A robust nonlinear mixed-effects model for COVID-19 deaths data. *Statistics and Its Interface*, **14**, 49–57.

Tsallis, C. and Tirnakli, U. (2020) Predicting COVID-19 peaks around the world. *Frontiers in Physics*, **8**, 217.

Verhulst, P. F. (1838) Notice sur la loi que la population suit dans son accroissement. *Corresp. Math. Phys.*, **10**, 113–126.

— (1844) *Recherches mathématiques sur la loi d'accroissement de la population*. Mémoires de l'Academie royale des sciences des lettres et des beaux-arts de Belgique.

— (1847) *Deuxième mémoire sur la loi d'accroissement de la population*. Bruxelles: Hayez.

Wang, X.-S., Wu, J. and Yang, Y. (2012) Richards model revisited: Validation by and application to infection dynamics. *Journal of Theoretical Biology*, **313**, 12–19.

Yamana, T. K., Kandula, S. and Shaman, J. (2016) Superensemble forecasts of dengue outbreaks. *Journal of The Royal Society Interface*, **13**.

4

Data distributions

Guido A. Moreira
Universidade Federal de Minas Gerais, Brazil

Juliana Freitas
Universidade Federal de Minas Gerais, Brazil

Dani Gamerman
Universidade Federal de Minas Gerais/Universidade Federal do Rio de Janeiro, Brazil

CONTENTS

This chapter introduces the probabilistic models used throughout this book. The focus here is on models for count data. The traditional Poisson model is presented as well as its potential drawback involving the relation between mean and variance of this distribution. Then, some possible alternatives to overcome this obstacle are presented. Finally, some important issues regarding the data modelling procedure are discussed.

DOI: 10.1201/9781003148883-4

4.1   Introduction

The growth curves (i.e., logistic, generalised logistic curve, among others)
presented in the last chapter, as well as other models for the number of cases
in an epidemic, usually account for the average, or expected value, of cases.
However, in many situations, modelling in this manner is not sufficient to
explain the variations of the data around this average. Furthermore, these
variations, also called residuals, generally do not present any recognisable
pattern, as can be seen in Figure 4.1.

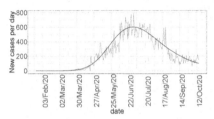
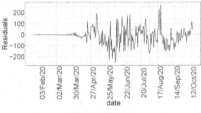

(a) Observed data along with estimated
mean curve.

(b) Residuals.

FIGURE 4.1: Data, model fitting and residuals of data of Nigeria. Panel 4.1a
shows the evolution of confirmed cases of COVID-19 in Nigeria, from 23 January 2020 to 31 October 2020 along with the fit of generalised logistic model.
Panel 4.1b concerns the residuals calculated as the observations minus the
fitted value at each time $t = 1, 2, \ldots, 262$. Data source: Database prepared by
the Center for Systems Science and Engineering at Johns Hopkins University.

Figure 4.1 shows the evolution of the confirmed cases of COVID-19 in
Nigeria between 23 January 2020 and 31 October 2020, along with the fit
of generalised logistic model to this data. Panel 4.1b presents the residuals,
calculated as the difference between observations and their respective fitted
value at each time $t = 1, 2, \ldots, 262$. Thus, negative values for the residual
indicate overestimation, while positive accounts for underestimation.

From a statistical perspective, the observed data can be considered as
being a variation around a mean. In other words, the exact observed counts are
described as a dispersion around the average under some probabilistic norm.
This book will cover the basic properties of random variables and distributions.
The reader interested in more details about this matter is referred to DeGroot
and Schervish (2012) and Ross (2018).

The probabilistic norm mentioned above is related to other concepts such
as event, random variable, mean, variance, and many others. A focus is given
on the ones that are important for the understanding of how count data is
treated in this book. Then, in slightly more detail, the random variable and the
probability for all possible events need to be well defined. In turn, the variance

or standard deviation, i.e., the square root of the variance, is a popular and useful measure of dispersion mentioned earlier. The standard deviation can be interpreted as the typical deviation of the data from the mean. However, a statistical model requires more than average and dispersion. Since a thorough take on probability will not be done, definitions are presented in informal style.

Definition 6 *(Random variable and event) A random variable is a function connecting all possible values of an experiment with a respective probability. An event is a grouping of possible values of a random variable which can be measured by probability.*

The most direct use of the random variable is through the quantification of the results of an experiment. Additionally, the set of all possible values of a random variable is denoted Ω. This event has probability 1.

Given the concepts described above, another important point in statistical modelling can be introduced: the probability distribution. A probability distribution is a function that associates events of a random variable to probabilities. Here, it is worth making a distinction between discrete and continuous random variables. This distinction relies mainly on Ω which has, respectively, countable and uncountable cardinalities. The essential consequence of this distinction is that the probability of events with only one value is necessarily zero for continuous random variables. In this case, infinitesimal increases of the probability are measured by a function called the (probability) density function, usually denoted $f(y)$. The probability of an event A is $\int_A f(y)dy$. Nonetheless, for a discrete random variable, say Y, a probability function $P(Y = y)$ is available and the probability of an event A is given by $\sum_{y \in A} P(Y = y)$.

Two important properties of a probability distribution are the expected value and variance. For a discrete random variable, these functions are defined, respectively, as:

$$E(Y) = \sum_{y \in \Omega} yP(Y = y) \text{ and } Var(Y) = \sum_{y \in \Omega} (y - E(Y))^2 P(Y = y). \quad (4.1)$$

If Y is considered to be continuous, the equations in (4.1) are now defined as

$$E(Y) = \int_\Omega yf(y)dy \text{ and } Var(Y) = \int_\Omega (y - E(Y))^2 f(y)dy. \quad (4.2)$$

It is important to highlight that, regardless of the type of the random variable (i.e., continuous or discrete), it is always true that $Var(Y) \geq 0$. Also, functions that estimate both the expected value and the variance are close to their respective theoretical values in some sense.

In what follows, more than one random variable can be dealt with at the same time. In addition, these different random variables can relate to one another, and this relationship can be characterised through their joint distribution.

The joint distribution quantifies the probabilities of simultaneous events of the random variables. Joint probabilities are denoted by separating the representation of these variables by a comma. For example, if X and Y are continuous random variables, then $f(x, y)$ represents their joint density. Still considering two random variables, there can be the situation where they are related to each other. This case can be thought of as a change of pattern of one variable depending on what happened with the other. Or, in an opposite scenario: the variables have no influence on each other. These ideas lead to the concept of independence. Two random variables are said to be independent if $f(x, y) = f(x)f(y)$ for the continuous case, or $P(X = x, Y = y) = P(x = x)P(Y = y)$ if the random variables are discrete. More about this definition can be found in Ross (2018).

These variables can also be related through conditional probabilities. Considering only two random variables, the idea of conditional probabilities is to quantify the probabilities of the events of one random variable *given* the knowledge of the realised value, or set of values, of the other. In order to represent conditional probabilities, a vertical bar is used. Thus, if X and Y are continuous random variables, $f(x \mid y)$ represents the conditional density of X given $Y = y$. Next, analogous to the definitions in Equations (4.1) and (4.2), the joint and conditional expectation and variance can be defined.

Using the concept of joint distribution and independence, two important results are introduced. They are the laws of total expectation and the law of total variance, determined in Equation (4.3):

$$
\begin{aligned}
E(X) &= E\left(E(X \mid Y)\right) \\
Var(X) &= E\left(Var(X \mid Y)\right) + Var\left(E(X \mid Y)\right).
\end{aligned}
\tag{4.3}
$$

With all these concepts in mind and a data set at hand, perhaps a primary objective in a statistical analysis is to find a distribution which is likely to have generated these data. It is usual to adopt a distribution indexed by a vector of parameters. This procedure tends to make the problem simpler, as finding a small dimensional parameter vector is easier than finding an entire distribution. The notation $Y \sim \mathscr{P}(\theta)$ means that the random variable Y follows a distribution $\mathscr{P}$ with parameter(s) θ.

A common assumption for data with multiple measurements is that they are conditionally independent given the parameters, which implies that

$$
P(Y_1 = y_1, \ldots, Y_m = y_m \mid \theta) = P(Y_1 = y_1 \mid \theta) \ldots P(Y_m = y_m \mid \theta).
\tag{4.4}
$$

This assumption often makes sense and it makes many calculations easier, as will be seen later in Section 4.4. Briefly, under the knowledge of the parameters θ, the information coming from Y_i does not affect the distribution of Y_j, for $i \neq j$.

This section concerned a few introductory concepts and properties of probability and distributions. These concepts are important as they will be extensively used throughout this book. In the next sections, examples of possible distributions associated with count data will be dealt with, as well as their main advantages and potential drawbacks.

4.2   The Poisson distribution

The Poisson distribution is possibly the first distribution that comes to mind when one thinks about count data. It is defined on the set $\Omega = \mathbb{N}$. The immediate implication of this definition is that, when a random variable Y is assumed to follow this distribution, its possible values are $y \in \{0, 1, 2, \dots\}$. This distribution is indexed by only one parameter, which is commonly denoted by λ. Here, the parameter λ can be interpreted as a rate; therefore, it is a positive real value. From a practical point of view, the concept of rate can be related to the magnitude of the outcomes: a higher rate indicates that the random variable Y may assume larger values; on the other hand, lower rate favours low counts. If it is indexed by time, it can also indicate temporal changes along the observation window. The probability function of the Poisson distribution is

$$P(Y = y) = \frac{e^{-\lambda}\lambda^y}{y!}, \text{ for } y = 0, 1, 2, \dots. \tag{4.5}$$

Then, assuming that $Y \sim Poisson(\lambda)$, the expected value and variance are given by

$$E(Y) = \lambda \text{ and } Var(Y) = \lambda. \tag{4.6}$$

That is, under this distribution, the expectation and the variance are equal.

The parameter λ plays an important role in the probability function as it determines the probabilities of observing each possible value y. Naturally, it dictates both the mean and the variance. In some situations, it makes sense to associate these probabilities with time. For example, consider the counts Y as the daily number of confirmed cases of a disease. It may be probable to observe 0 counts at the beginning of the follow-up; however, large counts are more likely when the peak number of cases is reached. In this situation, one can associate these data with the Poisson distribution and the function of time can be one of those listed in Chapter 3. These two quantities can be combined, by making λ a function of the time and a set of other parameters. In more detail, for each time t, one may have

$$y_t \mid \mu(t) \sim Poisson(\mu(t)), \text{ for } t > 0 \tag{4.7}$$

where y_t is the number of new cases in time t and $\mu(t)$ is the chosen curve for modelling the data. Thus, if the generalised logistic curve (Equation (3.7)) is used, the assumption $y_t \mid \mu(t)$ is equivalent to $y_t \mid (a, b, c, f)$, as of now the unknown parameters is the vector that composes the function $\mu(t)$, i.e., (a, b, c, f).

The property in Equation 4.6 of the Poisson distribution can be too restrictive, since the values of the variance are only tangible in conformation with the scale of the data. An exercise can be undertaken to clarify this issue. This exercise consists of simulating data using random number generators, which are available in any statistical software. The basic idea of this activity is to verify whether a proposed model typically generates data that resemble true observations, both concerning their pattern over time and their observational variation. This verification can be done via comparison metrics, or, in our simple example, in the form of visual comparison.

Consider real data of confirmed cases of COVID-19 from Brazil, Germany and Nigeria for this exercise. After obtaining the data, the model described in Equation (4.7) with $\mu(t)$ given by the generalised logistic model (see Equation (3.8) for more details) was fitted, whose procedure is described in Chapter 6. Table 4.1 shows the parameter estimates for the selected countries. Next, simulated values were generated based on the parameter estimates and $t = 1, 2, \ldots, 266$ for Nigeria, $t = 1, 2, \ldots, 233$ for Brazil and $t = 1, 2, \ldots, 113$ for Germany. In order to do so, the software R (R Core Team, 2020) version 4.0.3 with a seed of 123 was used. A comparison between the real data and data generated from the fitted model is displayed in Figure 4.2.

TABLE 4.1: Estimates of the vector of parameters of the generalised logistic growth model, considering the confirmed cases of Brazil, Germany, and Nigeria. Observations from 23 January 2020 to 15 October 2020 for Brazil and Nigeria, and from 23 January to 14 May for Germany. Data collected from the Center for Systems Science and Engineering at the Johns Hopkins University and the Brazilian Health Ministry.

Country	a	b	c	f
Brazil	7071.206	0.487	0.019	9.461
Germany	8.4×10^{-13}	0.037	0.084	12.253
Nigeria	0.043	0.091	0.026	5.957

The reader interested in understanding the role of each of the parameters in Table 4.1 can go to Sections 3.2 and 3.3.

Comparing the left panels with the right panels of Figure 4.2, it can be seen that there certainly are variations around the curve in the simulated data. Yet, these variations are not as large as the variation of the actual observations. This behaviour happens because the mean function was well accommodated to data, as can be observed in Panels 4.2c and 4.2a. Nevertheless, given that

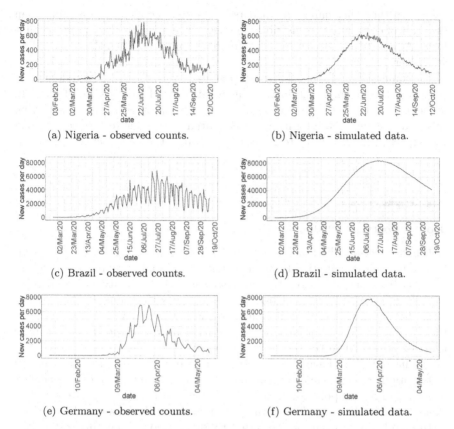

FIGURE 4.2: Comparison between observed counts of new daily confirmed cases and simulated data. Left panel: observed counts. Right panel: simulated data generated from the fitted model. Data time period in 2020: For Nigeria, from 23 January to 15 October. For Brazil, from 25 February to 15 October. For Germany, from 23 January to 14 May. Data source: database prepared by the Center for Systems Science and Engineering at Johns Hopkins University.

the variance is equal to the mean, the dispersion around the mean is shrunk, see Panels 4.2d and 4.2b.

A comparison of the countries' outcomes shows that the magnitude of the Brazil counts is higher than that of both Germany and Nigeria. It seems then that the simulated values based on the estimated parameters using Brazil counts are less disperse when compared to the fit to the observed counts of cases in Nigeria. In the latter, simulated data are less disperse in comparison. This happens precisely because the dispersion allowed by Poisson distribution, which is provided by the variance $\lambda = \mu(t)$, is tied to the very expected value $\lambda = \mu(t)$. Thus, when the measurement values grow, so does the expected value. On the other hand, the dispersion around the mean grows more than the scale of data, and this distribution does not manage to adequately account for it. As a result, this pattern is even more accentuated when the count values are larger.

In summary, in Figure 4.2, note that data variations are completely attached to the mean, which in consequence is related to the scale of data. In order to turn this situation around, authors have proposed ways of softening this relationship. These forms allow the variance to be lower or greater than the expectation. These strategies are discussed in the next section.

4.3   Overdispersion

In the last section, the Poisson distribution was introduced and it was shown how to use this distribution to model count data. In addition to that, a potential drawback of this distribution in what concerns flexibility and robustness was also presented. This drawback is the property that, under this distribution, the variance is necessarily equal to the mean, i.e., $Var(Y) = E(Y)$. In the present section, alternatives to the situations where the variance can be lower or greater than the mean will be discussed.

Count data that have greater dispersion than that allowed by the Poisson distribution are often called overdispersed data. Accounting for overdispersion usually signifies proposing a distribution for the data whose variance is greater than the expected value. This strategy can be applied in many different ways in the modelling procedure. An interesting discussion about this matter can be found in McCullagh and Nelder (1989).

When it comes to epidemic count data specifically, Chowell et al. (2019) seek to explain possible sources that lead to overdispersion. They mention that, in many cases, aggregated data of sub-regions are the source of the overdispersion. For example, the counts of a country are composed of a sum of the states (or other types of subdivision) of this country. Yet, if the analysis is on a singular state, this challenge may still occur as a state is a sum of municipalities with different characteristics and timings. The authors refer to this situation as multiple overlapping sub-epidemics. This issue will be readdressed in Chapter 5.

A straightforward way to extend the Poisson distribution to allow for overdispersion is by adding a hierarchy level. In other words, assume that $Y \mid \lambda \sim Poisson(\lambda)$ and that λ can be a random variable itself. Then, an interesting result comes from the law of total expectation and the law of total variance introduced in Section 4.1, Equation (4.3). This result is,

$$E(Y) \quad = \quad E\left(E(Y \mid \lambda)\right) = E(\lambda) \tag{4.8}$$

and

$$Var(Y) = Var\left(E(Y \mid \lambda)\right) + E\left(Var(Y \mid \lambda)\right)$$
$$= Var\left(E(Y \mid \lambda)\right) + E(\lambda)$$
$$\geq E(\lambda). \tag{4.9}$$

Note that the result in Equation (4.9) is true because $Var\left(E(Y \mid \lambda)\right) \geq 0$. This outcome is a consequence of the property that variances are always non-negative, as mentioned in Section 4.1. In addition to that, the random variable Y has larger variance compared to its expected value, which is the goal. In order to ensure that the epidemic data have the desired mean curve $\mu(t)$, all that is needed is to set $E(\lambda) = \mu(t)$. Recall that possible forms for $\mu(t)$ were discussed in Chapter 3.

Proceeding with the details of the inclusion of the overdispersion characteristic into a modelling approach, let us focus on the parameter λ. A common choice is to use $\lambda \sim Gamma(\alpha, \beta)$, where both α and β are positive real values. The probability density function of the Gamma distribution is

$$f(\lambda) = \frac{\beta^\alpha}{\Gamma(\alpha)} \lambda^{\alpha-1} e^{-\beta\lambda}, \lambda > 0, \tag{4.10}$$

where $\Gamma(\alpha) = \int_0^\infty t^{\alpha-1} e^{-t} dt$. The theoretical mean and variance for this distribution are

$$E(\lambda) = \frac{\alpha}{\beta} \text{ and } Var(\lambda) = \frac{\alpha}{\beta^2}. \tag{4.11}$$

An interesting characteristic of this distribution is that the vector of parameters (α, β) can be rewritten in terms of the mean and variance, leading to $\alpha = (E^2(\lambda)/Var(\lambda))$ and $\beta = E(\lambda)/Var(\lambda)$.

Once again, by using the laws of total expectation and total variance (Equation (4.3)), it is straightforward to derive the marginal theoretical mean and variance of Y. They are

$$E(Y) = E\left(E(Y \mid \lambda)\right) = E(\lambda) = \frac{\alpha}{\beta} \tag{4.12}$$

and

$$Var(Y) = Var\left(E(Y \mid \lambda)\right) + E\left(Var(Y \mid \lambda)\right)$$
$$= Var(\lambda) + E(\lambda) = \frac{\alpha}{\beta}\left(1 + \frac{1}{\beta}\right). \tag{4.13}$$

Note that the variance of Y is equal to its expected value times a component that is greater than one. Consequently, it is clear that variance is no longer equal to the mean. Another implication of this choice of distribution is the fact that, marginally, $Y \sim NB(\alpha, 1/(\beta + 1))$, where NB stands for Negative Binomial. Moreover, the component $(1 + (1/\beta))$ is called the overdispersion component. For clarity, the probability function of the Negative Binomial distribution in this case is given by

$$P(Y = y) = \frac{\Gamma(\alpha + y)}{\Gamma(\alpha)y!} \left(\frac{\beta}{1 + \beta}\right)^{\alpha} \left(\frac{1}{1 + \beta}\right)^{y}, \text{ for } y = 0, 1, \ldots$$

So far, all the discussion in this section has concerned a possible way to include overdispersion in the modelling process. Nonetheless, as was mentioned earlier, there is more than one form to treat this characteristic. Then, in Sections 4.3.1 and 4.3.2, two choices for the parameters α and β are provided, which, in turn, affect different features of the overdispersion.

4.3.1   Negative Binomial: Mean-dependent overdispersion

Growth curves that can be used to describe the pattern of counts of cases in an epidemic were discussed in Chapter 3. Moreover, this function can be associated with the mean of the count distribution. Then, since it is desired that the marginal mean for the data y_t is a growth curve $\mu(t)$, any parameterisation— of the Poisson, Negative Binomial or other distribution—should enforce this choice. Alternative representations of α and β in Equations (4.12) and (4.13) define an overdispersion parameter ϕ. In the alternative of this section, establish $\alpha = \phi$ and $\beta = \phi/\mu(t)$, so that

$$E(y_t) = \mu(t) \text{ and } Var(y_t) = \mu(t) \left(1 + \frac{\mu(t)}{\phi}\right). \tag{4.14}$$

In this case, the set of parameters includes those that index the mean function $\mu(t)$ and the overdispersion *parameter* ϕ. In addition to that, in Equation (4.14), the term $(1 + (\mu(t)/\phi))$ is the overdispersion *component*. The parameter ϕ controls how much overdispersion is present in the model. Thus, as ϕ decreases, the overdispersion increases. On the other hand, as $\phi \longrightarrow \infty$, the present model converges in distribution to a Poisson model with mean $\mu(t)$. This parameterisation is ideal when the overdispersion increases when the data gets larger.

The simulation exercise in Section 4.2 is repeated here. The idea is to generate data from the model and to evaluate whether a data set generated from this distribution resembles the true observations. Assume that the daily number y_t of new confirmed cases of COVID-19 in Germany followed a Negative Binomial distribution with the parametrisation described in Equation (4.14) above, with $\mu(t)$ given by the generalised logistic model (see Equation (3.8)). Next, parameter estimation was performed and finally, data was simulated with the same assumptions, based on the values of the estimated parameters. The data and results of the simulations are displayed in Figure 4.3.

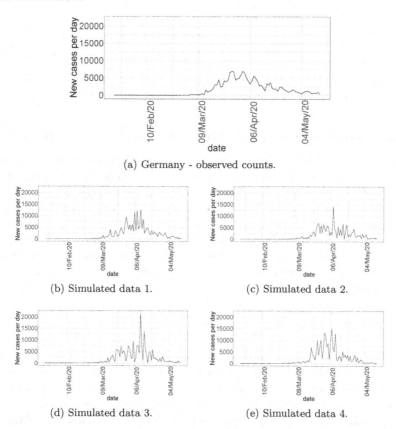

(a) Germany - observed counts.

(b) Simulated data 1.

(c) Simulated data 2.

(d) Simulated data 3.

(e) Simulated data 4.

FIGURE 4.3: Comparison between real data of new daily confirmed cases, model fitting and generated data. Panel 4.3a shows the real data from Germany, from 23 January 2020 to May 14, 2020. In turn, Panels 4.3b to 4.3e concern four simulated data based on the parameters of a fitted model and the mean-*dependent* parametrisation. Parameter estimates on which data generation was based were $a = 3.18 \times 10^6$, $b = 0.805$, $c = 6.415$, $f = 1.231$, and $\phi = 0.087$. Data source: database prepared by the Center for Systems Science and Engineering at Johns Hopkins University.

It can be clearly seen from Figure 4.3 that the model allows more dispersion that does the Poisson model, confirming the theoretical result previously described. Moreover, there are some outliers, mainly in Panels 4.3c and 4.3d. That feature is allowed by the Negative Binomial distribution. This feature is adequate with data coming from very different sources, in which spurious data occasionally appear; and it is frequently encountered in pandemic data.

A final point about this specific parameterisation of the Negative Binomial distribution is that it is a very intuitive one, in the sense that it is equivalent to the model

$$y_t \mid \mu(t), \lambda_t \sim Poisson(\mu(t)\lambda_t)$$
$$\lambda_t \sim Gamma(\phi, \phi). \tag{4.15}$$

For this reason above all, this is the most used parameterisation in practice. Still, useful alternative is presented in the next section.

4.3.2 Negative Binomial: Mean-independent overdispersion

In the previous subsection, the parametrisation used implied that the overdispersion component was given by $1 + \mu(t)/\phi$. Therefore, this component depends both on the mean function $\mu(t)$ and the overdispersion parameter ϕ. Thus, overdispersion decreases/increases along with the mean. Nonetheless, since the variance $Var(y_t) = \mu(t)\left(1 + \mu(t)/\phi\right)$ already increases with the expected value, this strategy may not always be a pertinent path to follow. Alternatively, one may want the overdispersion to not depend on the mean. This possibility is allowed by choosing $\alpha = \phi\mu(t)$ and $\beta = \phi$ in Equations (4.12) and (4.13). In this case,

$$E(y_t) = \mu(t) \text{ and } Var(y_t) = \mu(t)\left(1 + \frac{1}{\phi}\right). \tag{4.16}$$

Now, the overdispersion component in Equation (4.16) is $1 + 1/\phi$, and it does not depend on the mean, as intended. In the current case, ϕ also controls the amount of overdispersion in the model. Furthermore, it has an easy interpretation, as $1/\phi$ represents the percentage increase for the variance over the mean. Again, when $\phi \longrightarrow \infty$, the Poisson model (see Section 4.2) is obtained.

The previous simulation exercise is repeated one more time in order to illustrate this option of parametrisation. The results with the previous two versions of the model can be reviewed in Figures 4.2 and 4.3. The data and results of the simulations are displayed in Figure 4.4.

It can be seen from Figure 4.4 that all the four simulated data series resemble very closely the real data observations, from both the mean and the variance points of view. This similarity provides evidence that the Negative Binomial model with the current parametrisation might be adequate to fit these data. Finally, it is worth pointing out that these specifications made use of the versatility of the Gamma distribution in terms of the relation between its mean and variance.

In this subsection, an alternative parametrisation for the Negative Binomial distribution, the main characteristics, and properties are presented. Other characterisations are possible, including ones that maintain the coefficient of variation—i.e., standard deviation divided by the mean—for y_t constant over t. Then, the next section explores other possibilities for modelling the overdispersion without using the Negative Binomial distribution.

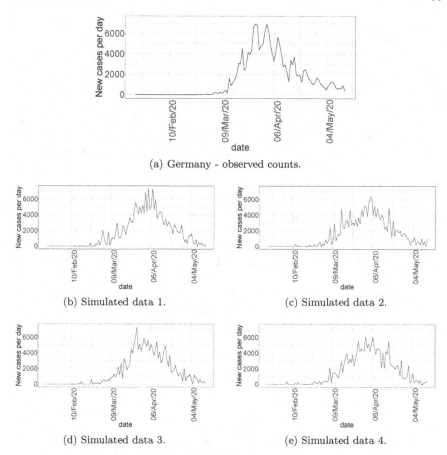

(a) Germany - observed counts.

(b) Simulated data 1.

(c) Simulated data 2.

(d) Simulated data 3.

(e) Simulated data 4.

FIGURE 4.4: Comparison between real data of new daily confirmed cases, model fitting and generated data. Panel 4.3a shows the real data from Germany, from 23 January 2020 to 14 May 2020. In turn, Panels 4.3b to 4.3e concerns four simulated data based on the parameters of a fitted model and the mean-*independent* parametrisation. Parameter estimates on which data generation were based are $a = 0.035$, $b = 0.001$, $c = 0.104$, $f = 2.289$, and $\phi = 2.407$. Data source: database prepared by the Center for Systems Science and Engineering at Johns Hopkins University.

4.3.3 Other models for the overdispersion

As mentioned in the beginning of this section, there are several ways to include overdispersion in a modelling process. Perhaps the most immediate one is the inclusion of a stochastic hierarchy, via introduction of an additional layer with a random component λ. In the previous section, this idea was dealt with by assuming that $\lambda \sim Gamma(\alpha, \beta)$, where the vector (α, β) could assume different parametrisations. As a result, the marginal distribution of y_t in this case was the Negative Binomial. After that, one obvious question is, what if one chooses another distribution for the parameter λ?

Quite often, different distributions for λ result in a situation in which the marginal distribution of y_t is unknown. Even so, it is still possible to evaluate its marginal expected value and theoretical variance by using the laws of total expectation and total variance (see Equation (4.3)). This affirmation leads one to conclude that whatever the choice for the distribution of λ is, only its expected value $\mu(t)$ must be retained. This assumption is based on previous knowledge of a likely overall temporal behaviour of the mean of the random variable y_t. This reasoning implies that $E(y_t) = \mu(t)$. Also note that the distribution for λ must be concentrated over positive values.

One immediate choice is the log-Normal distribution. This choice presents an interesting intuition: the cumulative number of cases $M(t)$ is usually close to a linear trend in the log scale, and the log-Normal distribution represents Gaussian variations around the mean in the same log scale. This characteristic is discussed in Chapter 3, see Figures 3.1 and 3.2. On the other hand, looking at the daily new confirmed cases of COVID-19 series of Argentina, Armenia, Japan, and Luxembourg, it seems that in some cases the overdispersion comes mostly from a delay in the notification associated with some weekdays. The data of the mentioned countries can be seen in Figure 4.5.

One way to try to model this feature is to use $\lambda = e^z$, where $z \sim SN(m, \sigma^2, \alpha)$, and SN stands for a Skew Normal distribution with (negative) skewness α. The aforementioned log-Normal case happens when $\alpha = 0$. Marginally, λ has log-SN distribution. The density function of this distribution is

$$f(\lambda) = \frac{2}{\sqrt{2\pi}\lambda\sigma} \exp\left\{-\frac{(\log \lambda - m)^2}{2\sigma^2}\right\} \Phi\left(\alpha\left(\frac{\log \lambda - m}{\sigma}\right)\right), \text{ for } \lambda > 0.$$

The expectation and the variance of λ are given, respectively, by

$$E(\lambda) = 2\exp\left(m + \frac{\sigma^2}{2}\right)\Phi\left(\frac{\sigma\alpha}{\sqrt{1+\alpha^2}}\right)$$

and

$$Var(\lambda) = 2\exp\left(2m + \sigma^2\right)\left(\exp\left(\sigma^2\right)\Phi\left(\frac{2\sigma\alpha}{\sqrt{1+\alpha^2}}\right) - 2\Phi\left(\frac{\sigma\alpha}{\sqrt{1+\alpha^2}}\right)^2\right),$$

where $\Phi(x) = \int_{-\infty}^{x} \frac{1}{\sqrt{2\pi}} e^{-\frac{x^2}{2}} dx$. So, choosing $m = \log\left(\frac{\mu(t)}{2\Phi\left(\frac{\sigma\alpha}{\sqrt{1+\alpha^2}}\right)}\right) -$

$\frac{\sigma^2}{2}$ implies that $E(\lambda) = \mu(t)$.

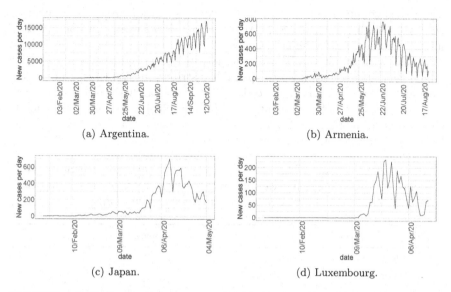

(a) Argentina.   (b) Armenia.

(c) Japan.   (d) Luxembourg.

FIGURE 4.5: Daily confirmed cases of COVID-19 in Argentina (Panel 4.5a), Armenia (Panel (4.5b), Japan (Panel 4.5c), and Luxembourg (Panel 4.5d). The starting date in all cases was 23 January 2020. The last evaluated day was 17 October, 25 August, 04 May, and 16 April of 2020, respectively. These series exhibit data whose dispersion comes mostly from sudden drops in reported cases. Data source: database prepared by the Center for Systems Science and Engineering at Johns Hopkins University.

As a consequence, by the law of total variance, the marginal variance of y_t is

$$Var(y_t) = \mu(t) \left\{ 1 + \mu(t) \left(\exp\left\{\sigma^2\right\} \frac{\Phi\left(2\sigma \frac{\alpha}{\sqrt{1+\alpha^2}}\right)}{2\Phi\left(\sigma \frac{\alpha}{\sqrt{1+\alpha^2}}\right)^2} - 1 \right) \right\}. \quad (4.17)$$

Note that $\left\{ 1 + \mu(t) \left(\exp\left\{\sigma^2\right\} \frac{\Phi\left(2\sigma \frac{\alpha}{\sqrt{1+\alpha^2}}\right)}{2\Phi\left(\sigma \frac{\alpha}{\sqrt{1+\alpha^2}}\right)^2} - 1 \right) \right\}$ is the overdispersion component, which is dependent on the mean. In the case of the log-normal distribution ($\alpha = 0$), the overdispersion reduces to $\left\{1 + \mu(t) \left(\exp\{\sigma^2\} - 1\right)\right\}$. In addition, the hyperparameter σ of the log-SN distribution has a similar interpretation to the hyperparameter ϕ of the Negative Binomial distribution of the previous section. Thus, any positive distribution can be used for λ.

Alternatively, the conditional model for y_t does not even have to be Poisson. The Poisson distribution is used since it comes naturally from the aggregated counts of a Poisson process. Also, the mean curve $\mu(t)$ integrates very easily with the λ parameter. However, any other model for non-negative integers can potentially be used instead.

In this section, alternative ways of modelling the overdispersion were presented. Hence, considering all the possible distributions and the modifications therein, there is a list of options to choose from. Certainly there might be more complex possibilities for modelling this data. Table 4.2 summarises the models considered in this chapter. Therefore, it also enables a clearer comparison between these different options.

TABLE 4.2: Summary of distribution models and their parametrisations to model count data. Mean and variance represent the marginal mean and variance of y_t. NB stands for the Negative Binomial distribution.

Dispersion model	Mean	Variance
Poisson (no overdispersion)	$\mu(t)$	$\mu(t)$
NB, mean-dependent	$\mu(t)$	$\mu(t)\left(1 + \dfrac{\mu(t)}{\phi}\right)$
NB, mean-independent	$\mu(t)$	$\mu(t)\left(1 + \dfrac{1}{\phi}\right)$
Log-Normal	$\mu(t)$	$\mu(t)\left\{1 + \mu(t)\left(e^{\sigma^2} - 1\right)\right\}$
Log-SN	$\mu(t)$	$\mu(t)\left\{1 + \mu(t)\left(e^{\sigma^2}\dfrac{\Phi\left(2\sigma\dfrac{\alpha}{\sqrt{1+\alpha^2}}\right)}{2\Phi\left(\sigma\dfrac{\alpha}{\sqrt{1+\alpha^2}}\right)^2} - 1\right)\right\}$

For all cases in Table 4.2, the distribution mean is $\mu(t)$. This pattern occurs to reflect the expected evolution in time of the counts of new cases of a disease. The next section brings important discussion concerning both data and specific modelling characteristics.

4.4   Discussion

In the previous sections, important probabilistic concepts and some of the inherited properties were introduced. The Poisson distribution and possible ways to accommodate the overdispersion characteristic were also considered.

Such ways included distributions like the Negative Binomial. Nonetheless, a few discussion points rise from the statistical modelling perspective. Thus, in order for the analysis to be reliable, all the assumptions must be well understood and enforced. One of the most important discussions is about which type of data should be considered. For example, should one consider the number of new cases or the cumulative number of cases at each time t? This choice has a proper answer based on the theory about the likelihood function, which, in turn, is the main tool for estimating models in Statistics. Therefore, Section 4.4.1 presents arguments to indicate the appropriate way to model the counts of cases in an epidemic. Focusing in another topic, Section 4.4.2 is about the possibility of truncating the range of a parameter. This possibility may be very useful, mainly when there is previous information on the amount of dispersion the model should present.

4.4.1 Daily new cases vs cumulative cases

The likelihood function in an important concept in the core of statistical analysis. It is a function of the parameters representing how likely a parameter value is in the light of the data information, and it is based on the joint probability for the data. In the previous sections, likelihood-related topics were used to describe the conditional distribution of the *data* given the parameters, for example. In practice, the parameters' vector are usually unknown. As a consequence of this lack of knowledge, the observed data is used to learn about the parameters. So, the likelihood function, denoted $l(\theta; y) = l(\theta)$, is written as

$$l(\theta; y) = l(\theta; y_1, \ldots, y_n) = f(y_1, \ldots, y_n \mid \theta), \tag{4.18}$$

meaning that it is numerically equal to the joint probability model for the data $y = (y_1, \ldots, y_n)$. According to the likelihood principle (Birnbaum, 1962), the likelihood function summarises all the information available from the data about the parameters' vector. For this reason, special care must be taken.

A common agreement about the likelihood function is that each single measurement from the data should be counted only once. This agreement is based on the intuitive idea that repeating information biases results and one should only assume to know what one knows; not more or less. Referring to counts of cases in an epidemic, by using the cumulative number of cases in each time t, one is carrying and repeating this information over and over again. For example, as $Y_2 = y_1 + y_2$, assuming the observed data to be (Y_1, Y_2) means that y_2 will be counted once, as it should, but y_1 will be counted twice. Proceeding with this reasoning, by time t, Y_t will be carrying repeated information from $t-1$ time points, and only in t the new one is aggregated. Thus, a solution to this matter is to consider the number of new cases at each time t. Other solutions based on cumulative counts could also be possible, but they require further assumptions (of conditional independence) that unnecessarily complicate the models.

Furthermore, an assumption is made that $y_t \sim \mathscr{P}$ with a mean given by $E(y_t) = \mu(t)$. Then, considering that $Y_t = \sum_{i=1}^{t} y_i$, the mean of Y_t should be given by the sum $\sum_{i=1}^{t} E(y_i) = \sum_{i=1}^{t} \mu(i)$. This is a slight modification over the model initially proposed for cumulative counts in Chapter 3. Comments about this modification were also introduced in the same chapter.

Another obstacle of using the cumulative counts Y_t is that the likelihood function is based on the joint distribution of the data given the parameters. The most direct way to obtain the joint probability of this function is to assume the data are conditionally independent given the parameters. However, if the data is composed of the cumulative daily counts, every day is necessarily equal or greater than the day before. This relationship implies that these data are not conditionally independent. It is way simpler and straightforward to build a model based on the daily new cases. In this case, a joint model is easily specified by the conditional independence property.

In summary, there are sound reasons to endorse the use of data in the form of the new number of cases. They are (i) using data information only once and (ii) the simplicity of assuming conditionally independent data to compose the likelihood function. In the next section, a discussion about the possibility of truncating the range of parameter values is presented.

4.4.2 Parameter truncation

The Negative Binomial options discussed in this chapter were applicable to COVID-19 data in many locations, particularly when the daily cases have passed their respective peak dates. It can be likewise useful in any epidemic data. However, the previous affirmation about the Negative Binomial distribution might not always be the most adequate approach when one considers data collected before the peak, which is often hard to acknowledge until a set of observations after this date arrives (Chowell et al., 2020). When it comes to projections, this obstacle gets even more challenging. In addition to that, it is very important to concentrate on the uncertainty specification about the resulting outcomes. Hence, although a better explanation and discussion on this matter is provided in Chapter 6, especially in Section 6.1.4, some attention is anticipated in the present subsection.

The uncertainty about the prediction is represented by an interval in which, conditional on the past data, the probability that the next to be known observation y_{t+1} is in this interval is $100p\%$. The value of p is usually greater than or equal to 0.95, e.g., $\{0.95, 0.99, 0.999\}$. Small uncertainty is represented by an interval with relatively low amplitude, while predictions carrying a large amount of uncertainty have intervals with large amplitudes. In this regard, the predictions that resulted by using either one of the Negative Binomial parameterisations present an extremely large uncertainty. This is shown in Figure 4.6.

Figure 4.6 illustrates prediction intervals based on different models for the data of the daily confirmed cases of COVID-19 in Italy. From this figure, it can

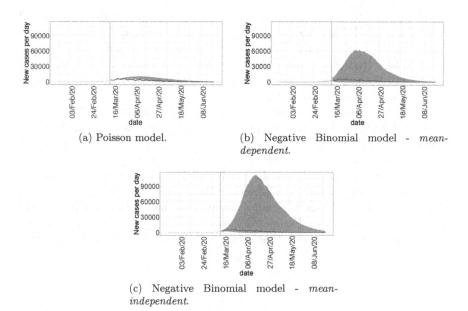

(a) Poisson model.

(b) Negative Binomial model - *mean-dependent*.

(c) Negative Binomial model - *mean-independent*.

FIGURE 4.6: Evolution of the daily confirmed cases of COVID-19 in Italy, from 23 January 2020 to 13 March 2020. This last date is previous to the peak. After this referenced date, the 95% prediction intervals based on three different approaches is presented in grey. Panel 4.6a refers to the Poisson model. In turn, Panels 4.6b and 4.6c concern the Negative Binomial model, in the *mean dependent* parametrisation and in the *mean independent* parametrisation respectively. Data source: database prepared by the Center for Systems Science and Engineering at Johns Hopkins University.

be noticed that the intervals in Panels 4.6b and 4.6c are considerably larger when compared to the one in Panel 4.6a. As a conclusion to this comparison, a dilemma involving the choice of the modelling approach appears. On one hand, the Poisson does not have enough dispersion for the data, as detailed in Section 4.2. On the other hand, when using the Negative Binomial model, projected cases seem to be very uncertain in the early stages. This very large uncertainty may make the decision-making process difficult.

This figure also shows that the uncertainty about the predictions from the Negative Binomial models is larger than the Poisson. However, by comparing the Poisson intervals with the realised future data, the prediction "misses" the data, albeit by a relatively small amount. In this sense, it might be more reasonable to have somewhat larger uncertainty in prediction intervals to include the future realisations.

These ideas naturally lead to the consideration of a trade-off between the Poisson and the Negative Binomial models. Note that, for both NB parametrisations, the lower the parameter ϕ is, the more overdispersion there is in the model. In an opposite scenario, as $\phi \longrightarrow \infty$, the Negative Binomial model converges to the Poisson model. For this reason, by limiting this parameter to be greater than a particular value it might be feasible to find an in-between model. This alternative would ideally present overdispersion, but with less uncertainty in the projections. Details about how this truncation is effectively done is discussed in Section 6.1.2.

It might be mentioned in advance that finding a good truncation value is not a straightforward procedure. This setting may be easier for the Negative Binomial case where the overdispersion component does not depend on the mean (see Equation (4.16)). This facility is due to the fact that, in this case, the overdispersion parameter ϕ is interpretable as a percentage increase of the variance with relation to the mean. However, when the overdispersion does depend on the mean, it is not so immediate, since the lower bound will itself depend on the mean. Figure 4.7 shows the same projections based on the count data of COVID-19 cases that occurred in Italy using different truncation points for ϕ.

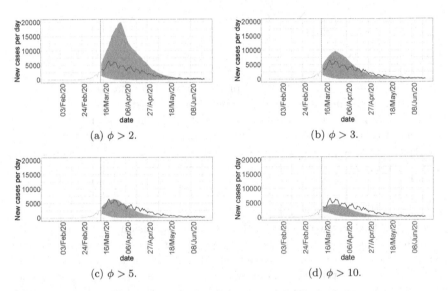

FIGURE 4.7: Prediction intervals of 95% for the confirmed cases of COVID-19 in Italy. Model is based on the Negative Binomial with the mean-dependent parametrisation and truncating the overdispersion parameter. Four different values for the truncation were considered. Panel 4.7a shows results with $\phi > 2$; Panel 4.7b illustrates the case of $\phi > 3$; Panel 4.7c concerns $\phi > 5$; and finally, Panel 4.7d targets the range $\phi > 10$. Data source: database prepared by the Center for Systems Science and Engineering at Johns Hopkins University.

It can be seen that some projections look closer to the Poisson model while others are more similar to the unrestricted Negative Binomial model.

Therefore, it is indeed possible to choose a model "between" the two options: Poisson and Negative Binomial. That is, a model that is able to accommodate the overdispersion characteristic. Still, a compromise needs to be achieved between too much and too little dispersion or uncertainty. Adding uncertainty leads to wider intervals which are more likely to include the future data, but may end up being too uncertain to be useful. An additional challenge presents itself since a truncation point for one set of data may result in different characteristics for another set. So, fine tuning may be required for the desired result.

Bibliography

Birnbaum, A. (1962) On the foundations of statistical inference. *Journal of the American Statistical Association*, **57**, 269–306.

Chowell, G., Luo, R., Sun, K., Roosa, K., Tariq, A. and Viboud, C. (2020) Real-time forecasting of epidemic trajectories using computational dynamic ensembles. *Epidemics*, **30**, 100379.

Chowell, G., Tariq, A. and Hyman, J. M. (2019) A novel sub-epidemic modeling framework for short-term forecasting epidemic waves. *BMC Medicine*, **17**.

DeGroot, M. H. and Schervish, M. J. (2012) *Probability and Statistics*. Boston: Pearson, 4th edn.

McCullagh, P. and Nelder, J. A. (1989) *Generalized Linear Models*. New York: Chapman and Hall/CRC, 2nd edn.

R Core Team (2020) *R: A Language and Environment for Statistical Computing*. R Foundation for Statistical Computing, Vienna, Austria. URLhttps://www.R-project.org/.

Ross, S. M. (2018) *A First Course in Probability*. Boston: Pearson, 10th edn.

5

Modelling specific data features

Guido A. Moreira
Universidade Federal de Minas Gerais, Brazil

Juliana Freitas
Universidade Federal de Minas Gerais, Brazil

Leonardo Nascimento
Universidade Federal do Amazonas, Brazil

Ricardo C. Pedroso
Universidade Federal de Minas Gerais, Brazil

CONTENTS

Epidemics or pandemics datasets may present several different specific features. These features, in turn, may play an important role in the particular case being studied. Then, it may be of interest to accommodate such characteristics in the modelling process as it can enhance the understanding of the disease as well as to result in more accurate projections, for example. Modelling challenges involve handling heterogeneous sub regions nested in a larger region of interest, seasonality induced in the data by the notification date, and also presence of multiple waves of outbreak. Considering this framework, the present chapter is composed of descriptions and modelling strategies that can be used in each of the circumstances mentioned above. An additional comment is that the reader may experience a different set of challenges in the epidemic of her/his interest. The approach presented here to tackle these specific issues could hopefully serve as inspiration to generate modelling ideas on how to handle other data specific issues that might emerge.

DOI: 10.1201/9781003148883-5

5.1 Introduction

Chapters 2 and 4 discussed some characteristics that can be found in pandemic data, such as different concepts of cases between different countries, low availability of tests, delays in notification, over-dispersion and under-dispersion. However, there are other characteristics present in pandemic data and the current chapter will introduce modelling tools to handle some of them.

Section 5.2 handles the heterogeneity present in the data, that is, the propagation of the pandemic happens differently in each predetermined sub-region. This characteristic is easier to observe in countries with a large territorial extension, for example Brazil. Figure 5.1 shows that the Brazilian states have different stages of the pandemic for the same period. In this case, the region is Brazil and the sub-regions are its states. Another important data feature, presented in Section 5.3, is seasonality. Basically, this behaviour occurs when there is, repeatedly, a decrease or increase in notifications on specific days of the week. Figures 12.9a and 12.10 show this pattern for Brazil, Costa Rica and Paraguay. In particular, this section will focus on situations where there is only a decrease in notifications. Finally, Section 5.4 presents the situation of multiple waves. In this case, a wave is an upward trend to the peak followed by a downward trend in notifications after the peak. Consequently, the feature "multiple waves" is the repeated presence of a this pattern. Figure 5.8 shows an example of multiples waves. The statistical models presented in this chapter are mostly based on Equation (3.8) to illustrate how to accommodate these specific characteristics.

5.2 Heterogeneity across sub-regions

The current stage of the spread of an ongoing pandemic and its future predictions are commonly analysed through data sets aggregated at national or state/province/county levels. As will be discussed in Chapter 9, daily counts of new cases and new deaths due to the COVID-19 pandemic were made available by most of the countries in the world. These data sets are usually provided at a national level, but sometimes also at the state and city levels.

Some factors may affect the spread dynamics of a pandemic in a specific region. For example, regions with greater demographic densities may lead to higher transmission rates and, therefore, to the occurrence of a higher number of deaths. On the other hand, a collective effort of a region's population to prevent the disease dissemination can save human lives. Appropriate hygiene measures in addition to strict social distance policies have been shown to be effective practices to reduce the spread of the disease. These and other relevant factors, such as the available economic resources and the quality of the healthcare system, may occur in opposite directions within different sub-regions of a larger region, as states and cities of a country, for example. One consequence is that it may generate heterogeneous patterns of the pandemic course across the different sub-regions.

The pandemic can be in distinct stages within its states and cities, especially in a large country. A decline stage in the counts of new cases or deaths in a group of states, where the disease cases first appeared, may occur concurrently with a growth stage of these counts in another group of states, where the disease cases started later.

Figure 5.1 presents the moving averages of the daily counts of new deaths by COVID-19 for all 27 Brazilian states. The Paraná (PR), Rio Grande do Sul (RS) and Santa Catarina (SC) states belong to the south region of Brazil. Their counts follow similar growing patterns over time for the range displayed in the figure. This is a case of homogeneity across states. On the other hand, it is clear that the disease appeared in Ceará (CE) and São Paulo (SP) (northeast and southeast regions, respectively) before it occurred in the southern states. It is also noticeable that the counts in CE started to decline while the counts in the southern states had just started to grow. These different stages of the pandemic between CE and the southern states represent an example of heterogeneity across sub-regions of a larger region. SP has the greatest number of deaths among all the Brazilian states. This position can be somewhat expected as this is the state with the largest population in the country.

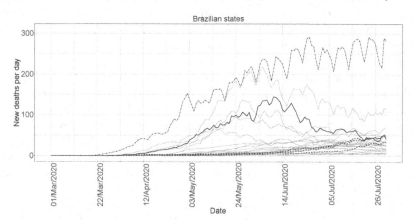

FIGURE 5.1: Moving average of range nine of the daily new deaths by COVID-19 in Brazilian states (gray solid lines). The highlighted curves are from Ceará (solid black line), São Paulo (dash-dotted black line) and the southern states of Paraná, Rio Grande do Sul and Santa Catarina (dashed black lines). The time period is from 1st March 2020 to 31st July 2020. Data source: Brazilian Ministry of Health.

If there is no heterogeneity within a group of sub-regions, unique time-dependent model (the generalised logistic function, for example) may be suitably assumed to model the aggregate counts of the group. In the cases where there is heterogeneity among the sub-regions of a larger region, a specific model should be assumed for each sub-region. Inference about the full region should be performed by aggregating the estimates of the models of each sub-region, instead of estimating unique growth curve model for the data aggregated over the entire region. For example, based on the model presented in Equation (4.7), assume that $y_{i,t} \mid \mu_i(t) \sim Poisson(\mu_i(t))$, for $i = 1, \ldots, d, t > 0$, where $y_{i,t}$ is the number of new deaths (or confirmed cases) by a disease in the sub-region i at day t, $\mu_i(t)$ is any time-dependent function appropriate to model the shape of the counts in state i, and d is the total number of sub-regions that form a partition of some specified larger region. Define the aggregate count by $y_t = \sum_{i=1}^{d} y_i(t)$. It follows that

$$y_t \mid \mu_1(t), \ldots, \mu_d(t) \sim Poisson\left(\mu(t)\right), t > 0, \tag{5.1}$$

where $\mu(t) = \sum_{i=1}^{d} \mu_i(t)$. The mean and variance estimates for $\mu(t)$ are equal to the sum of these respective estimates for $\mu_1(t), \ldots, \mu_d(t)$. The median and credible intervals for $\mu(t)$ are not equal to the sum of the related quantiles estimated for each sub-region $i = 1, \ldots, d$. If the estimation of the models for each sub-region is made through some of the MCMC methods presented in Section 6.2.2, a sample $\mu^{(s)}(t)$ from $\mu(t) \mid (y_1, \ldots, y_t)$ is provided by the set of values $\mu^{(s)}(t) = \sum_{i=1}^{d} \mu_i^{(s)}(t)$, where s is the iteration index of each sample of $(\mu_1(t), \ldots, \mu_d(t))$. This way, the median and the credible intervals for $\mu(t)$ may also be estimated. The same procedure described above for the *Poisson* model may be adapted to other probability models, for example, the ones presented in Table 4.2.

Figure 5.2 presents the estimates of $\mu(t)$ and the predicted values for the COVID-19 death counts in Brazil considering a single model for the aggregate count of the country and the combined estimates of the models for each state. The same data presented in Figure 5.1 is considered for the observations. A single generalised logistic growth model is assumed for $\mu(t)$ for the aggregated count and for each $\mu_i(t)$, $i = 1, \ldots, 27$, corresponding to the state-level counts. The MCMC procedure to compute the estimates displayed in Figure 5.2 will be detailed in Chapter 10.

The decreasing predicted values for the single model are undesirable because many Brazilian states present a growth trend of the death counts, as shown in Figure 5.1. The combination of models provides predicted values that are compatible with the many different trends among the analysed states, indicating the possibility of increasing or decreasing curves for the future aggregate death counts in Brazil.

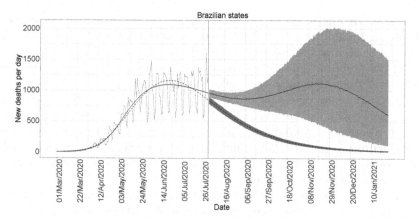

FIGURE 5.2: Observed aggregate death counts in Brazil from 1st March 2020 to 31st July 2020 (gray solid line), median estimates for $\mu(t)$ for the single model (dashed black line) and for the combination of models for each state (solid black line), with respective 95% credibility intervals for the predicted values (gray shadows).

5.3 Seasonality

Another important characteristic of the data that must be considered in this chapter is seasonality. This feature could also be referred to as reporting delay, or even as the "weekend effect". So, in this section this topic is targeted and possible forms of modelling are discussed.

Seasonal effects may occur when data are released by the notification date. Then, at this point, it is important to have in mind the distinction between notification or reporting date and other disease-related dates. Based on the rationale of the timing of the disease, one can think of (i) the date of first symptom, (ii) date of a medical appointment, (iii) date of testing for the disease, (iv) date of the test result, (iv) date when the case is entered into the health system, among others (for more details, see Marinović et al. (2015)). The latter item is called the reporting date, as previously explained in Chapter 2. For several possible reasons, the period of time between dates (i) and (iv) may be small, but could also be days and months apart.

Focusing on the reporting date, it may depend on a number of reasons, as well as the health staff delay in effectively registering this information in the health system. For example, during weekends and holidays, there might be a reduced capacity for performing this task. Also, lower counts may occur on specific weekdays as a simple consequence of having less people working on the respective days immediately before. It is noteworthy that these days may vary from region to region, as well as the number of days presenting this pattern of "camouflaged" lower counts. Some of these ideas were discussed in Chapter 2. In addition, a form of detecting and evaluating this feature in the data can be seen in Chapters 12 and 13.

All the procedures described above are reflected in the series of reported events over time. Generally, these effects cause a pattern of consistent decays in the counts on specific days of the week. After such days, the scale of the counts is resumed. Figure 5.3 illustrates the situation being discussed. As time evolves, a consistent pattern of an increase in the number of cases followed by a sudden drop is observed. What happens next is a return to the previous data levels, or even counts maintaining values at a higher level. This characteristic is present in the series exhibited in the figure.

An idea for dealing with this situation is to include a parameter representing the information that, on a specific day of the week or on a few specific days, lower case counts are expected. Then, let $l \in \{1, 2, \dots, 7\}$ be the index representing a day of the week. For instance, if $l = 1$, then the time t refers to a Sunday; in turn, when $l = 6$, the date represented by t was a Friday. Moreover, define a parameter d_l and a set $\mathbf{D}_l = \{t \text{ is a day corresponding to day } l \text{ of the week}\}$, for $l = 1, 2, \dots, 7$. The strategy presented here consists of multiplying this parameter by the mean curve for a day of the week with lower counts (due to the delay in the reporting). In this way, a fraction of the mean $\mu(t)$ will be accounted for that selected day. Also, note that if $d_l = 1$ for some l, the ordinary model with no effect on the delay in the reporting is obtained.

In what follows, assume there is a reporting delay associated with day l of the week. If $t \in \mathbf{D}_l$, the mean of the number of new cases in time t is decreased by $100d_l\%$, i.e.,

$$E(y_t) = \mu^*(t) = \begin{cases} \mu(t)\, d_l, & \text{if } t \in \mathbf{D}_l \\ \mu(t), & \text{if } t \notin \mathbf{D}_l, \end{cases} \tag{5.2}$$

for $t = 1, 2, \dots, n$. Alternatively, it is possible to rewrite Equation (5.2) as

$$\begin{aligned} E(y_t) &= d_l\mu(t)1\{t \in \mathbf{D}_l\} + \mu(t)1\{t \notin \mathbf{D}_l\} \\ &= \mu(t)\left[d_l 1\{t \in \mathbf{D}_l\} + 1\{t \notin \mathbf{D}_l\}\right] \end{aligned} \tag{5.3}$$

with $1\{.\}$ being the indicator function. From Equations (5.2) and (5.3) it is possible to see that the mean curve for the days that do not present a delay in the notification is the same as the one introduced and explored in previous chapters. However, when t corresponds to the l-th day of the week, the mean curve is only the fraction d_l of $\mu(.)$.

A possible extension of this approach is to include this effect for more than one day of the week. This extension is quite straightforward: one needs simply to add a term $d_{l'}\mu(t)$ for $t \in \mathbf{D}_{l'}$, where l' is another day of the week, different from l. In practice, the total number of days with this departure from the

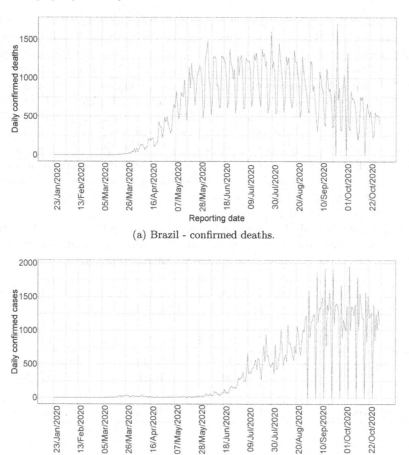

(a) Brazil - confirmed deaths.

(b) Costa Rica - confirmed cases.

FIGURE 5.3: Evolution of the confirmed deaths of COVID-19 in Brazil (Panel 5.3a) and the confirmed cases registered in Costa Rica (Panel 5.3b), per notification date. The period of evaluation varied from 23/January/2020 to 31/October/2020. Data source: Brazilian Ministry of Health and Center for Systems Science and Engineering at Johns Hopkins University.

standard model should not be over 2 or at most 3 to avoid poor identification of the main curve.

The model specification is completed with a distribution for y_t, $t = 1, 2, \ldots, n$, a definition for the growth curve $\mu(.)$ and for the day(s) having this effect, as well as prior distributions for the additional unknown quantities. This information is discussed in the preceding chapters of Part II and in Chapters 12 and 13. As an example, the model described in Equation (5.2) was fitted for the daily number of confirmed deaths from COVID-19 in Brazil and for the daily confirmed cases in Costa Rica (see raw counts in Figure 5.3). In this example, a Poisson distribution was assumed for the data and the generalised logistic was chosen for the mean. In addition, the days that presented a delay in the reporting were Sunday and Monday for Brazil, and only Sunday for Costa Rica. Also, projections for the future outcomes were computed based on definitions described in Chapter 6. The results are displayed in Figure 5.4.

The estimated mean curve $\mu^*(.)$ is highlighted in both panels of Figure 5.4. They take into account the information that the data indicate a lower count of cases on one (Panel 5.4b) or two (Panel 5.4a) days of the week. Furthermore, it is clear that this curve provides a better fit to the data pattern in the sense that it closely follows the observations than a model without these seasonal effects. The regular mean curve is not able to present such a feature without the addition of seasonal components.

Following the analysis, additional interpretation can be obtained from the estimation of seasonal parameters d_l. Table 5.1 displays point and interval estimates for these parameters to enable this understanding.

TABLE 5.1: Point and interval estimates based on the posterior distributions of the parameters representing the decay in the reporting pattern.

Country	Parameter	Median	Standard Deviation	HPD 95%
Brazil	d_1	0.492	0.005	[0.483, 0.500]
	d_2	0.559	0.005	[0.549, 0.568]
Costa Rica	d_1	0.304	0.004	[0.295, 0.313]

First, considering Brazil, the posterior median of d_1 was 0.492 with an HPD interval of $[0.483, 0.500]$. This means that it is expected that only 49.2% of the cases are actually reported on Sundays, and this percentage is compared against the number of cases that would be accounted for if there were no delays in the reporting. Roughly speaking, on average, only approximately half of cases are in fact reported on Sundays in Brazil. This conclusion is similar to Monday, in which 55.9% of the cases were effectively reported in this day. The interpretation of the "weekend effect" in Costa Rica follows the pattern observed for Brazil. On Sundays, the number of reported cases is expected to be only 30.4% of the regular amount.

Several modifications can be applied to the previous ideas. For example, if one does not know if the departure from the regular pattern is lower or greater than the mean of the other days, it is possible to loosen the range of d_l. This range was previously considered to be $(0, 1)$; now, d_l could vary in $(0, \infty)$. In this case, if the posterior density of d_l is concentrated in $(0, 1)$, it means that on that day, data indicate a reduction over the otherwise expected count means. In opposition, if the posterior probability for d_l is concentrated in the region

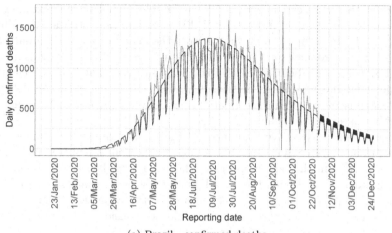

(a) Brazil - confirmed deaths.

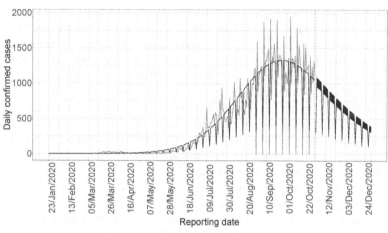

(b) Costa Rica - confirmed cases.

FIGURE 5.4: Data series (grey solid line) along with model fit (black solid line) accommodating the "weekend effect" and projections (grey area) for the confirmed deaths of COVID-19 in Brazil (Panel 5.4a) and the confirmed cases notified in Costa Rica (Panel 5.4b). The vertical dotted line indicates a separation between observed counts and projections. The raw counts alone can be seen in Figure 5.3. Data source: Brazilian Ministry of Health and Center for Systems Science and Engineering at Johns Hopkins University.

$d_l > 1$, the day presents, on average, a greater amount of reporting compared to its expected pattern. Finally, when the value 1 falls within the credibility limits, that day of the week can be assumed as following the same mean pattern; in this case, one can assume no "special effect" on that day. The model including the "weekend effect" is already implemented in the `PandemicLP` package, to be described in Part VI of this book and in de Freitas Magalhães et al. (2021). In addition, note that neither the peak nor the ending date will change. Thus, properties described in Section 3.3 are still valid for this model. One last comment about modelling the seasonality is that it is not treating the delay in the reporting; that is, data may still come with the delay. If the reader is interested in the relationship between the occurrence and the reporting dates, go to Chapter 7.

The presentation above contemplates the most common and simplest approach to handle weekly seasonality. By assuming that only a few days of the week depart from the regular pattern, the remaining days of the week were left unchanged, simplifying the model. One might argue that counts for these remaining days would need to compensate for the departure from standard for the chosen days. The proposal above did not consider this compensation, possibly because the standard for the remainder would have already been adapted.

There is no simple way to accommodate the variation above in the model. The most obvious one is to allow for the complete weekly pattern with effects d_l, for all weekdays, $l = 1, ..., 7$. These effects are not jointly identified unless some restriction is imposed on them. In an additive model, one of them or their total sum could be fixed to 0. In the multiplicative approach used here, this would translate into assuming one of the d_l's to be 1. The main problem is that it is not obvious which day should be set as having no seasonal effect. The alternative in this situation is to assume the effects sum to 0 (in an additive model) which translates in the present case to setting the d_l's to satisfy $\prod_l d_l = 1$. This deterministic seasonal model is more complete, richer and hence more computationally demanding.

All the comments above apply only to situations where the seasonality is fully deterministic and does not change over time. It is not uncommon in the time series context to observe seasonal patterns with such changes. In this case, a simple alternative, in line with Section 8.3, is to accommodate a dynamic seasonality. There are many ways of implementing it. As a simple example, this could involve replacing the assumed knowledge of the product of seasonal effects to equal 1 by a more flexible option $\prod_{t=j}^{j+6} d_t = w_t$, where the disturbances w_t have mean 1 and some positive variance. The role played by these disturbances is to allow for temporal variation of seasonal effects. The static, deterministic seasonal model is obtained in the limiting case where all disturbances w_t are degenerated with variance 0, i.e., they are all concentrated in the value 1, for all t. The reader should visit Section 8.3 for more details on dynamic models and their application to pandemic modelling.

So far, the only type of seasonality considered has a weekly cycle. Of course, there is no reason to restrict seasonal effects on pandemic counts to weekly effects. Porojnicu et al. (2007), Hayes (2010), and Juzeniene et al. (2010) are just a small sample of documented evidence regarding the influence of other forms of seasonality in the outcome of epidemics. These consist of those typically associated with the climate and the annual weather cycle. Prata et al. (2020), Rouen et al. (2020), Xie and Zhu (2020) attempted to associate the growth of COVID-19 cases with environmental-related factors, such as temperature and humidity. Once again, if there is knowledge about which variables are causing the seasonal effect, they could be included in the model following steps to be provided in Section 8.2. If there is no clear evidence of which explanatory variables to include, then the seasonal effects presented above could be adapted.

At last, it should also be mentioned that, alternatively, one may simply group data in a less refined scale, e.g., week. This suggestion presents advantages and drawbacks. Grouping data may lead to a loss of information, since one would only be able to rely on the projections per aggregated window. In this case, the exactness of the daily information would no longer be available. Considering the situation of a very infectious pattern of transmission, projections may quickly become obsolete (Hsieh and Cheng, 2006; Chowell et al., 2020). In this case, it can be more interesting to keep data as daily counts and work with the weekend effect model discussed in this section. On the other hand, if grouping data does not bring any severe loss of information or one is interested in studying the pattern of notifications, one may use the strategy of grouping data and adopt other modelling approaches. Another point to consider is the amount of data available: in one month, there will be approximately 30 observations in a daily scale, but only 4 data points in the weekly scale. Then, having in mind the importance of projections to serve as a guide for studies or authorities, one can take into account the points of view presented here in each case-specific analysis. Chapter 13 presents a brief comparison of regular and weekly effect modelling daily counts and with regular modelling weekly counts.

5.4 Multiple waves

Sometimes pandemic data will not behave as expected by the growth curves discussed in Chapter 3. More specifically, after the peak, it is expected that the average number of cases only decreases over time, but this may not always happen. Figure 5.5 shows data for COVID-19 daily new cases, which proved the need to adjust the model.

By being a data-driven option, this book does not propose to explain why multiple waves happen or how the epidemiological concepts discussed in Chapter 3 relate to them. Instead, ways to model $\mu(t)$ so that they contemplate such behaviour are discussed.

In the literature, authors have been modelling two-wave data series in different ways. Wang et al. (2012) applied the Richards (or generalised logistic, see Equation (3.8)) model both to the H1N1 and the SARS outbreak

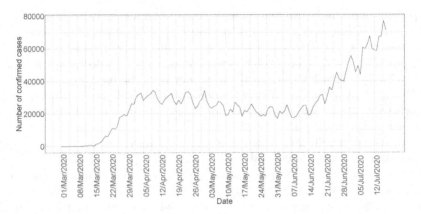

FIGURE 5.5: US COVID-19 daily cases from 1st of March to 18th of July, 2020 to illustrate the evidence of the appearance of a second wave.

in Canada, during the years 2009 and 2003, respectively. Although there is not an extensive description of their method, and it was not their main goal, it seems that they separated data into two new series concerning the waves. After that, an one-wave regular model can be used. In turn, Chowell et al. (2019) developed a more sophisticated approach. In brief, they introduced a random variable representing the cumulative number of cases at the end of the first wave. Consequently, the remaining cases would be placed in the second part of the epidemic. In other words, there is a separation of the stages in the epidemic, which is computed randomly. Despite these strategies, this chapter will discuss other ways to model the multiple wave behaviour in data series of epidemics.

There are countless ways to adapt the generalised logistic mean from Equation (3.8) to achieve two-curve representation. The proposals presented here change the means of the counts to contemplate more than one wave or curve. The model is Poisson using this mean as the rate parameter for each observation date, as discussed in Chapter 4. The most intuitive way to do this is to define:

$$\mu(t) = \mu_1(t) + \mu_2(t), \text{ where}$$
$$\mu_j(t) = \frac{a_j c_j f_j \exp(-c_j t)}{[b_j + \exp(-c_j t)]^{(f_j+1)}}, \; j = 1, 2. \tag{5.4}$$

Unfortunately, this proposal fails to capture the two waves. What has been verified is that either $\mu_1(t)$ or $\mu_2(t)$ is estimated at zero over the observed times and $\mu(t)$ has the same result as a single-wave model. As alternatives, the following options have been shown to work reasonably well, in the sense of providing good fit to the data.

The first option is a transition between the two curves:

$$\mu(t) \;=\; (1 - p(t))\mu_1(t) + p(t)\mu_2(t) \tag{5.5}$$

where $p(t)$ is a continuous probability distribution function, i.e., it is a continuous function, and $\lim_{t \to -\infty} p(t) = 0$ and $\lim_{t \to \infty} p(t) = 1$. A good option for the weight $p(t)$ is the Normal distribution function $\Phi\left(\frac{t-\eta}{\sigma}\right)$, where $\Phi(x) = \int_{-\infty}^{x} \frac{1}{\sqrt{2\pi}} e^{-\frac{u^2}{2}} du$. For this choice, parameter η represents the point where the second curve starts having larger weight than the first curve in the overall mean. Parameter σ indicates how fast the transition is, that is, larger values of σ indicate a smoother transition. The resulting fit for the US data can be seen in Figure 5.6.

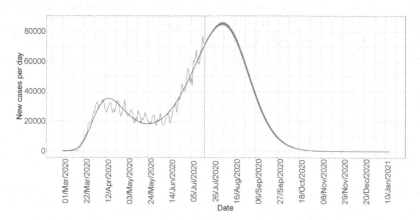

FIGURE 5.6: Fit of a smooth transition between two generalised logistic curves to the US data from 1st of March to 18th of July, 2020.

The fit in Figure 5.6 seems adequate for the data. Another similar option may be slightly more versatile. Consider

$$\mu(t) \;=\; \mu_1(t) + \mu_2(t), \text{ where} \tag{5.6}$$

$$\mu_j(t) \;=\; \frac{a_j c_j \exp(-c_j t)}{[b_j + \exp(-c_j t)]^2} \Phi(\alpha_j(t - \delta_j)), \; j = 1, 2, \tag{5.7}$$

where $\Phi(x)$ is the probit function. The difference in Equation (5.7) with respect to (5.4) is that the symmetric logistic curve ($f = 1$) is used and the skewness is induced by the $\Phi(\alpha(t - \delta))$ terms. The form for the skewness is motivated by the generalised class of skewed distributions in Azzalini (1985).

The versatility stems from the fact that the skewness is induced only if parameter δ_j is close to the jth curve peak. If it is not, then it can control the weight of a symmetric curve in $\mu(t)$, meaning that it can increase the weight of the jth curve around time δ_j abruptly or smoothly, depending on parameter α_j. The δ_j parameter can be "encouraged" to be near the peak of j's curve through its prior distribution. Instead of an independent prior, as frequently used, another choice can be made so that

$$E(\delta_j \mid b_j, c_j) = -\frac{\log(b_j)}{c_j}.$$

Note that $-\log(b_j)/c_j$ is the peak of the jth curve, as seen in Equation (3.10). The prior variance for δ_j can be set low or high, depending on how much the analyst wishes to force δ_j to be near the jth peak, which implies skewness. Additionally, the parameter α controls this skewness. Analogously to the truncation $f > 1$, it should be driven to be positive when the count data are expected to decrease after the jth peak slower than they increase before it. The means to drive the parameter values to be positive will be discussed in Section 6.1.2. The resulting fit is seen in Figure 5.7. The fit from Figure 5.7 also seems to be a good option for two-wave modelling and arguably better than the one from Figure 5.6. However, both options have been shown to be viable.

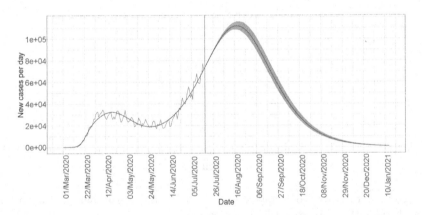

FIGURE 5.7: Fit of the sum of two logistic curves with probit induced skewness to the US COVID-19 data from 1st of March to 18th of July, 2020.

An advantage of the second proposal from Equation (5.7) is when the data shows evidence of more than two curves. As an example, see the COVID-19 data for the US up until November 1st, 2020 in Figure 5.8.

In the case of more than two waves, the model proposed in Equation (5.5) begins to get more complicated to apply as multiple transitions become necessary. On the other hand, the model proposed in Equations (5.6) and (5.7) presents a natural way to add curves. This model specification is presented below as

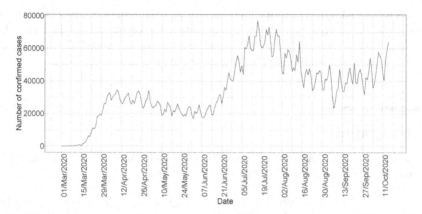

FIGURE 5.8: US COVID-19 daily cases from 1ˢᵗ of March to 16ᵗʰ of October, 2020 to illustrate the evidence of appearance of a third wave.

$$\mu(t) = \sum_{j=1}^{J} \mu_j(t), \text{ where}$$

$$\mu_j(t) = \frac{a_j c_j \exp(-c_j t)}{[b_j + \exp(-c_j t)]^2} \Phi(\alpha_j(t - \delta_j)), \; j = 1, \ldots, J,$$

where J is the desired number of waves or curves. The fit with $J = 3$ for the US data up to November 1 can be seen in Figure 5.9. The fit of the triple-wave model to the US data is adequate, giving reasonable predictions.

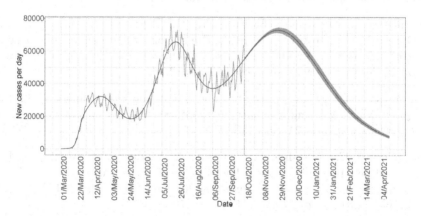

FIGURE 5.9: Fit of the sum of three logistic curves with probit-induced skewness to the US COVID-19 data from 1ˢᵗ of March to 16ᵗʰ of October, 2020.

As more and more components representing each individual wave are added to the model, the complexity and computational cost increases. However, the methodology presented here is still valid and can be used with the number of curves J as large as necessary to represent this feature of the data.

Bibliography

Azzalini, A. (1985) A class of distributions which includes the normal ones. *Scandinavian Journal of Statistics*, **12**, 171–178.

Chowell, G., Luo, R., Sun, K., Roosa, K., Tariq, A. and Viboud, C. (2020) Real-time forecasting of epidemic trajectories using computational dynamic ensembles. *Epidemics*, **30**, 100379.

Chowell, G., Tariq, A. and Hyman, J. M. (2019) A novel sub-epidemic modeling framework for short-term forecasting epidemic waves. *BMC Medicine*, **17**.

de Freitas Magalhães, D., da Costa, M. C. C. B., Moreira, G. A. and Menezes, T. P. (2021) *PandemicLP: Long Term Prediction for Epidemic and Pandemic Data*. URLhttps://CRAN.R-project.org/package=PandemicLP. R package version 0.2.1.

Hayes, D. P. (2010) Influenza pandemics, solar activity cycles, and vitamin D. *Medical Hypotheses*, **74**, 831–834.

Hsieh, Y.-H. and Cheng, Y.-S. (2006) Real-time forecast of multiphase outbreak. *Emerging Infectious Diseases*, **12**, 122–127.

Juzeniene, A., Ma, L.-W., Kwitniewski, M., Polev, G. A., Lagunova, Z., Dahlback, A. and Moan, J. (2010) The seasonality of pandemic and non-pandemic influenzas: The roles of solar radiation and vitamin D. *International Journal of Infectious Diseases*, **14**, e1099–e1105.

Marinović, A. B., Swaan, C., van Steenbergen, J. and Kretzschmar, M. (2015) Quantifying reporting timeliness to improve outbreak control. *Emerging Infectious Diseases*, **21**, 209–216.

Porojnicu, A. C., Robsahm, T. E., Dahlback, A., Berg, J. P., Christiani, D., Øyvind Sverre Bruland and Moan, J. (2007) Seasonal and geographical variations in lung cancer prognosis in Norway: Does vitamin D from the sun play a role? *Lung Cancer*, **55**, 263–270.

Prata, D. N., Rodrigues, W. and Bermejo, P. H. (2020) Temperature significantly changes COVID-19 transmission in (sub)tropical cities of Brazil. *Science of The Total Environment*, **729**, 138862.

Rouen, A., Adda, J., Roy, O., Rogers, E. and Lévy, P. (2020) COVID-19: Relationship between atmospheric temperature and daily new cases growth rate. *Epidemiology and Infection*, **148**, e184.

Wang, X.-S., Wu, J. and Yang, Y. (2012) Richards model revisited: Validation by and application to infection dynamics. *Journal of Theoretical Biology*, **313**, 12–19.

Xie, J. and Zhu, Y. (2020) Association between ambient temperature and COVID-19 infection in 122 cities from China. *Science of the Total Environment*, **724**, 138201.

6

Review of Bayesian inference

Guido A. Moreira
Universidade Federal de Minas Gerais, Brazil

Ricardo C. Pedroso
Universidade Federal de Minas Gerais, Brazil

Dani Gamerman
Universidade Federal de Minas Gerais/Universidade Federal do Rio de Janeiro, Brazil

CONTENTS

In this chapter, the theory and the operationalisation techniques for the models discussed in the previous chapters are presented. The focus is on Bayesian theory and its interpretation. Some modelling issues raised in Chapters 3 and 4 are addressed in more detail and the results from Chapter 5 can be better understood. Bayesian inference in most current problems typically implies integrating on distributions which are not known in closed form and require approximations. There are a few options relying on analytical and numerical techniques to handle the required calculations. Some of those are presented and discussed in the context of pandemic models. This chapter is meant to review some important concepts and results, but could also be used to introduce the basics of Bayesian inference to a beginner.

DOI: 10.1201/9781003148883-6

6.1   Inference

Solving problems with Statistics means associating probability distributions to observed data and unknown components, often being a parameter (vector) or future observations or both, in a mathematical model as discussed in Section 4.1, and making inference about these elements. In this context, this means aggregating and processing all information about the unknown quantities of such a model.

The most important source of information for a given set of parameters in a statistical model is the observed data. According to the likelihood principle (Birnbaum, 1962), all information from the data pertinent to a model's unknown parameters is carried by the likelihood function, which is defined on the parameter space. It is calculated as the joint probability (density) function for the data model evaluated at the observations and has different values for varying values of the parameters. If the data are modelled with a discrete distribution, the likelihood function can be loosely interpreted by the probability, for each parameter vector value, that a random draw from the data model would result in the measured observations.

For example, consider the Poisson model with the symmetrical logistic growth curve for the daily confirmed cases. In short, the model in this example is

$$y_t \mid a, b, c \quad \sim \quad Poisson(\mu(t)), \; t = 1, \ldots, n \tag{6.1}$$

$$\mu(t) \quad = \quad a\,c\,\frac{\exp\{-ct\}}{(b + \exp\{-ct\})^2}. \tag{6.2}$$

The unknown quantities of this model are (a, b, c). Another assumption of the model is the conditional independence of the data given the parameters, which facilitates building the joint model. Under this assumption, the joint probability function is factored into the product of n Poisson probability functions. Therefore, the likelihood function is

$$l(a, b, c; y) = p(y_1, \ldots, y_n \mid a, b, c) = \prod_{t=1}^{n} p(y_t \mid a, b, c)$$
$$= \prod_{t=1}^{n} e^{-\mu(t)} \frac{\mu(t)^{y_t}}{y_t!}. \tag{6.3}$$

The expression for mean function $\mu(\cdot)$ was not written in terms of the parameter vector (a, b, c) as in Equation (6.2) for simplicity. Note that once the data $y_1, \ldots, y_n$ are observed, Equation (6.3) is a function of a, b and c. By fixing one of the parameters, the two-dimensional function can be visualised through a contour plot. This type of two-dimensional figure shows a bivariate function. The varying colours show the function increase or decrease and the contours show regions with constant function values. The three contour plots of the likelihood in the log scale can be seen in Figure 6.1.

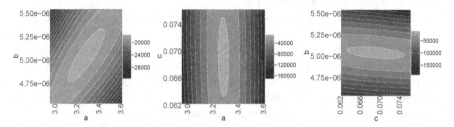

FIGURE 6.1: Contour plots of the likelihood function in the log scale for the Poisson model with logistic growth curve applied to the South Africa counts of COVID-19 cases.

In the particular case of the logistic model for the South Africa COVID-19 data in Figure 6.1, the conditional likelihood of a and b indicates a positive, possibly close to linear, relationship between these parameters. The other two pairs, however, seem nearly unrelated. It is common with the logistic and generalised logistic choices for the mean counts that some of the parameters are related in terms of the likelihood function.

One practical manner to aggregate information about the model parameters is to note that there is uncertainty about them. By associating a joint probability distribution to all quantities to which there is uncertainty, a very simple and straightforward procedure to make inference becomes apparent. This is called Bayesian inference. It is described in detail in Migon et al. (2014) and Gelman et al. (2013), but a short summary is presented here.

6.1.1 Bayesian inference

Denote by θ the p-dimensional vector of unknown quantities in a statistical model, also called the parameter vector, and by y the relevant data. The space of all possible values for θ is called the parameter space and is denoted by Θ. Then, Bayes' theorem can be used to combine any prior knowledge about θ in tandem with the information from the data through

$$\pi(\theta \mid y) = \frac{l(\theta; y)\pi(\theta)}{p(y)}. \tag{6.4}$$

The term $l(\theta; y)$ represents the likelihood function and it is the joint (density) probability function of the data, viewed as a function of the parameters (vector), as discussed above. Note that the likelihood function needs to be positive like a density function is, but there are no restrictions on its integral over Θ. The integral may even be infinity. The likelihood function is often interpreted as the information that the data set carries about the particular model. $\pi(\theta)$ is called the prior distribution and is a marginal distribution on θ that carries the information about the parameters which is known prior to the data collection. This is a very powerful tool in many ways, and its choice is briefly discussed in Section 6.1.2.

The resulting $\pi(\theta \mid y)$ is called the posterior density, or the density of the posterior distribution. It is called a posterior probability function if θ is discrete, but the term density is used here since all parameters in the discussed models are continuous. This density combines the information about the parameters brought by the data through the likelihood and brought by the analyst's opinion expressed through the prior distribution. As a result, this distribution is used to make inference and to drive decision making. A direct consequence of inference based on the posterior is the fact that the updated uncertainty about the parameter vector can be obtained from it.

The denominator $p(y) = \int_\Theta l(\theta; y)\pi(\theta)d\theta$ is commonly called marginal density, marginal likelihood. As θ is integrated out for the derivation of $p(y)$, it is only a constant in the expression of the posterior density. In this case, it guarantees that the posterior distribution integrates to 1, as it should. However, $p(y)$ can also be used for model comparison and selection through the Bayes factor (Bernardo and Smith, 2000, Section 6.1.4).

6.1.2   Prior distribution

When using Bayesian inference, choosing a prior distribution is just as important as choosing a model for the data. The prior carries all information known by the analyst and must be chosen with caution. This topic has been extensively discussed and the reader is referred to O'Hagan et al. (2006) and references therein. This section briefly discusses it and uses the prior choices made in the CovidLP app (http://est.ufmg.br/covidlp) and in its respective `PandemicLP` package illustrated in Chapter 15.

In many ways, the prior is one of the distinct features that makes Bayesian inference special. As long as the resulting posterior distribution is proper, which usually means that its integral does not diverge, the prior can be any positive function. This is always true when $\pi(\theta)$ is a probability distribution on Θ. The prior can be used to strengthen model estimation when there is little information in the likelihood function or to treat topological pathologies in the parameter space induced by the likelihood function (Gelman et al., 2017).

For very simple problems, some prior choices might be easily selected based on the model adopted for the data. This happens, for example, when the data follows a particular member of the exponential family, and is called conjugacy. For example, suppose $x_1, \ldots, x_n$ follow a Poisson distribution with common mean parameter θ. If the prior for θ is specified as a $Gamma(\alpha, \beta)$, then the posterior for θ is also a Gamma distribution, with parameters $\alpha + \sum_t x_t$ and $\beta + n$. Thus the Gamma is the conjugate prior for the Poisson model when the mean parameter is the same for all observations; see Migon et al. (2014)

for more details. However, most problems are more complex and cannot be simply solved with conjugacy.

The prior can also be used to add information known about the parameters, such as restrictions on the parameter space. For example, it might be common that the counts of pandemic cases and deaths increase over time faster than they decrease after the peak. This means that the parameter f in Equation (3.2) is greater than 1. This information can be included via the prior.

In particular, any information about the parameter space can be introduced in a "hard" way or in a "soft" way. The hard way is based on the fact that the posterior distribution is absolutely continuous with respect to the prior distribution (Schervish, 1995). Thus, regions of the parameter space could be given zero probability. Then, almost surely with respect to the marginal likelihood of the data, any region with zero probability in the prior will have zero probability in the posterior. On the other hand, the "soft" way consists of choosing a prior such that a small amount of probability is allocated to regions of the parameter space that are known to be unlikely. Note that the posterior distribution will probably not have the same small probability in that region, since the prior information is combined with the information from the likelihood.

As an important example, note that it is reasonable to expect the parameter f in the generalised logistic growth model to be larger than 1. There is a reasonable justification to include prior information for that, particularly in the period before the peak of cases. Until the number of daily cases begin to diminish, sometimes the information from the data carried by the likelihood function gives non-negligible weight to values of f smaller than 1. Since the behaviour of the data might indicate a slower descent than ascent, this information could be included in the prior. Sometimes, the probability weight for values smaller than 1 could be strong enough so that f might have small posterior probability to be greater than 1. This effect can be offset by using a truncated distribution where the prior probability for values of f below 1 is zero.

Another important truncation to be made is about the total number of cases. One of the characteristics of the generalised logistic mean curve for the Poisson model is that it produces unrealistically large predictions in the beginning of a pandemic, when there are few cases per day. This is verified by the total number of cases a/b^f sometimes being larger than that location's total population. By setting a truncation so that this value is smaller than a given number, say a percentage of the location's population, this explosive behaviour is avoided. One way to do this is to set a truncation so that $a/b^f \leq p\, N_{pop}$, where N_{pop} is the location's total population size and p is the chosen percentage. The default value for p used in the CovidLP app, for example, is 0.08 for cases and 0.02 for deaths.

A particular aspect of the generalised logistic mean model is that the parameters vector $\theta = (a, b, c, f)$ has some identifiability problems. Namely, there are many parameter quadruples that lead to the same curve, or at least very similar curves. This represents a pathology in the parameter space induced by the likelihood function, which can be treated by the prior. As an illustration of the information contained in the prior and the effect of the added information from the data through the likelihood function, the South Africa example is used again. Figure 6.2 shows the update of information for one parameter.

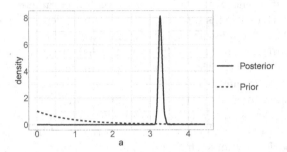

FIGURE 6.2: Comparison of the marginal prior and posterior densities for parameter a obtained from fitting the South Africa data. Prior: $Gamma(1, 1)$ distribution. Posterior: Obtained with methods described in Section 6.2.

The prior distribution provides little information about parameter a, as evidenced by a more "spread out" density. So, most of the information for the posterior comes from the data, that is, the likelihood function. It is interesting to see in Figure 6.2 that the prior has little information. On the other hand, the posterior is very informative, which represents the added value of the likelihood function. This means that the data has strong information towards the parameter in this model.

6.1.3   Estimation

An important feature of statistical inference is the estimation of the unknown quantities in the model. This is usually obtained via point or interval estimation or hypothesis testing. These topics are particular applications related to decision theory; see Robert (2007), Chapters 2 and 5. The reasoning for this is that estimation can be viewed as some summary of the posterior distribution. By being a summary of the distribution, some measure of information loss is assumed.

The loss of information is quantified by a loss function. So, for a point estimator, the estimated value is that which minimises the expected loss of the choice. Namely, the estimator for θ, denoted by $\hat{\theta}$ is the decision d for which the expected loss is the minimum, i.e.,

$$\hat{\theta} = \arg\min_{d} E\left[L(\theta, d) \mid y\right], \tag{6.5}$$

where $L(\theta, d)$ represents the loss of deciding for d when the true value of the parameter's vector is θ. As can be expected, changing the loss function changes the estimator. Popular choices of loss functions and their respective resulting estimators are seen in Table 6.1. In this case, $1_A(x)$ represents the indicator function which takes the value 1 when $x \in A$ and 0 otherwise.

TABLE 6.1: Common loss functions and their respective point-wise estimators

$L(\theta, d)$	Estimator
$(d - \theta)^2$	$E[\theta \mid y]$
$\lvert d - \theta \rvert$	Posterior median
$1 - 1_{\{\theta\}}(d)$	Posterior mode

Similarly, interval estimation also assigns a loss function for every possible interval of Θ. In this case, this can be reduced to choosing a region $\mathcal{C}$ such that:

$$P(\theta \in \mathcal{C} \mid y) \geq 1 - \alpha, \tag{6.6}$$

where α is an arbitrarily small value. A common choice is $\alpha = 0.05$. The interval $\mathcal{C}$ is also known as a credibility interval or region. Note that an infinite number of choices for $\mathcal{C}$ exist, regardless of α. For $\alpha = 5\%$, for example, $\mathcal{C} = (q_{0.05}, \infty)$ and $\mathcal{C} = (-\infty, q_{0.95})$, where q_p represents the posterior p^{th} quantile, are valid choices. However, if the idea of the estimation is to be as precise as possible, these intervals of infinite lengths are not very useful.

An optimised choice for the credible interval is the highest posterior density (HPD) interval, which can be equivalently defined as the interval of shortest length. In many cases, it is easier to obtain the equal tail probability interval given in the above example by $\mathcal{C} = (q_{0.025}, q_{0.975})$. This choice has an added benefit that if the posterior density is symmetric and unimodal with respect to its mean, then this choice is numerically identical to the HPD interval.

In the case of a model for an epidemic, it is possible to obtain estimates for the model parameters. Additionally, since $\mu(t)$ is a function of the parameters, the value $E(\mu(t) \mid y)$ can be evaluated as a point estimator for the expected value of cases at time t. This is the consequence of a quadratic loss for the estimation of the mean count. An interval can be likewise constructed, encoding the posterior uncertainty about $\mu(t)$.

In the case of hypothesis testing, the parameter space Θ is partitioned into k subsets $\{\Theta_j, j = 1, ..., k\}$. The posterior probability $P(\Theta_j \mid y)$ of each subset can be evaluated and a loss L_j is associated with each subset. The accepted hypothesis is the one with the smallest expected loss, that is, the one for which $L_j P(\Theta_j \mid y)$ is minimum. This topic is not further discussed here since hypotheses are not much tested throughout this book.

Coming back to the South Africa example, parameter a has a posterior distribution which can be visualised in Figure 6.3. The distribution is the same in Figure 6.2, zoomed in the region that the posterior distribution is more concentrated. Figure 6.3 displays the posterior mean for a, which is the point estimate with the quadratic loss. It also displays a 95% credibility interval. Its interpretation is very simple. The posterior probability that the true value for a is in the interval is 0.95.

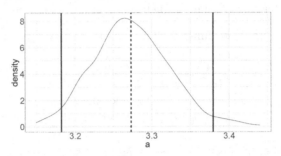

FIGURE 6.3: Marginal posterior density for parameter a obtained from fitting the South Africa COVID-19 data. Vertical lines are point (dashed line) and interval (solid lines) summaries of the distribution. The values are: Mean = 3.2744; 95% credibility interval = (3.1813, 3.3756).

6.1.4   Prediction

In order to forecast future observations z, a predictive distribution needs to be derived. For this purpose, it is useful to note that, in many cases, the observed data y consists of components that are conditionally independent given the parameter vector. Hence, future observations are also independent of past ones given the parameters. Therefore, the predictive distribution $p(z \mid y)$ is given by

$$p(z \mid y) = \int_\Theta p(z, \theta \mid y) d\theta = \int_\Theta p(z \mid y, \theta) \pi(\theta \mid y) d\theta. \qquad (6.7)$$

However, if z is conditionally independent of y given θ, then $p(z \mid y, \theta) = p(z \mid \theta)$. This is useful because $p(z \mid \theta)$ usually comes from the same family of distribution as the model for the observed data y given θ. Since $\pi(\theta \mid y)$ is the posterior density, the predictive density can be rewritten as

$$p(z \mid y) = E_{\theta \mid y} \left(p(z \mid \theta) \right), \qquad (6.8)$$

where $E_{\theta \mid y}(\cdot)$ represents the expected value taken with respect to the distribution of $\theta \mid y$, i.e., the posterior distribution. An advantage of the predictive distribution, from Equation (6.8), is that it carries all the uncertainty about the future observations, just as the posterior distribution does for the parameter vector.

    As a consequence, the whole discussion of estimation in Section 6.1.3 can also be applied to prediction, i.e., the predictive distribution. Therefore, point predictors can be obtained through the choice of a loss function, and prediction intervals can be constructed encoding the full uncertainty for the prediction.

An example, which was seen in Figure 4.7b and repeated in Figure 6.4, becomes clearer as the prediction interval for the future is plotted, encoding the uncertainty. Note that there is a prediction for each time point starting where the data stopped being collected, meaning that the prediction intervals are actually vertical in the plot. The grey shaded area is limited by the two lines connecting the upper limits and the lower limits of the marginal prediction intervals. These intervals mean that, for each time point t, there is a 95% predictive probability that the counts at that specific time are between the lines. However, it does not mean that the average line of future counts has a joint 95% predictive probability to be between the two plotted lines.

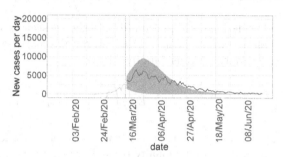

FIGURE 6.4: Example of a prediction interval for the predictive distribution, applied to data from Italy until 13 March 2020.

Unlike model parameters, predictions can be compared with their realised values since they can be observed once they are measured. A note needs to be made about prediction accuracy metrics which can be used for model comparison, as models with greater accuracy are preferred. Migon et al. (2014) argue that a point-wise comparison, such as the mean squared error of the realised value with respect to the predictive mean $E(z \mid y)$, is less ideal than comparing this value with a prediction that accounts for its full uncertainty, sometimes called probabilistic prediction or forecast. The Bayesian framework always provides forecasts through models that fully contemplate uncertainty. For the interested reader, probabilistic forecasts are discussed in detail by Gneiting et al. (2007) and Dawid (1991).

A straightforward idea to measure the prediction accuracy is to check how the realised value fits in the predictive distribution from Equation (6.8), meaning the value of the predictive density evaluated at the observed point. In particular, this is done in the log scale for a few different reasons such as computational stability, relationship with the Gaussian distribution, and relationship with the Kullback-Leibler information measure (Gelman et al., 2013). Then, if $z_1 = y_{n+1}$ is the observed value, $\log p(z_1 \mid y)$ is a prediction accuracy metric.

When there are multiple, say T, predicted points that have been observed, their joint predictive density can also be easily derived using the fact that they are conditionally independent given the parameters:

$$p\left(z_1, \ldots, z_T \mid y\right) = \int_\Theta p\left(z_1, \ldots, z_T, \theta \mid y\right) d\theta$$

$$= \int_\Theta p\left(z_1, \ldots, z_T \mid y, \theta\right) \pi(\theta \mid y) d\theta$$

$$= \int_\Theta \pi(\theta \mid y) \prod_{i=1}^{T} p\left(z_i \mid \theta\right) d\theta \tag{6.9}$$

$$= E_{\theta|y} \left(\prod_{i=1}^{T} p(z_i \mid \theta) \right).$$

The accuracy metric derived in Equation (6.9) in the log scale is an established and well-defined model comparison criteria. But note from Jensen's inequality that the log transformation needs care in this case:

$$\log p\left(z_1, \ldots, z_T \mid y\right) \geq E_{\theta|y} \left(\sum_{i=1}^{T} \log p(z_i \mid \theta) \right), \tag{6.10}$$

where the inequality indicates that the two sides of Equation (6.10) are not equal. Therefore, if computationally possible, the joint predictive density should be calculated in the natural scale, then integrated, then transformed to the log scale.

6.2　Operationalisation

There is a large variety of probability models that can be assumed for the observations and for the prior distributions, depending on the available data and the specific formulation and goals of each problem. This implies that many, possibly infinite, different configurations for the posterior distribution may be produced. An important definition in Bayesian inference is the class of conjugate families of distributions, already introduced in Section 6.1.2. A class $\mathscr{P}(\theta)$ of probability distributions is said to be a conjugate family with respect to some specific family $\mathscr{F}(\theta)$ of observational distributions if, for all observational distributions in $\mathscr{F}(\theta)$, the choice of a prior $\pi(\theta) \in \mathscr{P}(\theta)$ implies that $\pi(\theta \mid y) \in \mathscr{P}(\theta)$. This concept is also useful for some approximating techniques described later in this section.

As explained in Section 6.1.1, the denominator $p(y)$ in Eq. (6.4) is constant with respect to θ. It means that all the information about θ in the posterior distribution is provided by the likelihood and the prior distribution. The proportionality relation, with respect to θ, given by

$$p(\theta \mid y) \propto l(\theta; y) \, \pi(\theta) \tag{6.11}$$

is a relevant analytical aspect of the Bayes theorem. It allows the computation of $p(\theta \mid y)$, up to a multiplicative constant, regardless of the calculation of $p(y)$, which cannot be made analytically and has a high computational cost in many applications. The *Poisson-Gamma* conjugate family, exemplified in Section 6.1.2, is easily verified by the use of Equation (6.11). A model with likelihood $l(\theta; y) = e^{-n\theta} \theta^{\sum y_t} / (\Pi y_t!)$ and prior $\pi(\theta) = (\beta^{\alpha}/\Gamma(\alpha))\theta^{\alpha-1}e^{-\beta\theta}$ has a posterior distribution given by

$$p(\theta \mid y) \propto \theta^{\alpha + \sum y - 1} e^{-(\beta+n)\theta} . \qquad (6.12)$$

The expression in the right side of the proportional relation in (6.12) is the kernel of the $Gamma(\alpha + \sum y, \beta + n)$ density. It is the portion of the density function that depends on θ. This element determines the expression of the posterior distribution. Note that $p(y)$ is not relevant for the calculation.

As discussed in Section 6.1.2, the prior distribution may insert into the model more or less information about the parameter. The absence of prior information about θ may also be considered through non-informative priors. For example, the prior $\pi(\theta) \propto 1$, $\theta \in (0, +\infty)$, for the $Poisson(\theta)$ model considers that all values in $(0, +\infty)$ are, *a priori*, equally likely to be the true value of θ. This choice leads to a posterior distribution $Gamma(\sum_t y_t + 1, n)$, that is entirely specified using only information from the data. There exists a large variety of classes of conjugate prior distributions and different methods to specify a non-informative prior. The reader can find details about these topics in Migon et al. (2014) and Gelman et al. (2013).

Although conjugate families are appropriate to many applications, most of the real problems require the use of observational and prior distributions that do not result in analytically tractable posterior distributions. In these cases, the computation of the inference tools presented in Section 6.1 cannot be made exactly, and approximating methods are required. Some of these methods are briefly presented in the next sections. A more extensive presentation of approximating methods may be found in Migon et al. (2014).

Approximations must be performed in the cases when the posterior and predictive distributions are not analytically tractable to perform exact inference. In particular, inference can be summarised based on the features discussed in Sections 6.1.3 and 6.1.4. In general, desired features can be viewed as $E(h(\theta) \mid y)$ for an appropriate choice of $h(\cdot)$. This expectation is an integral, namely $\int_{\Theta} h(\theta)\pi(\theta \mid y)d\theta$. For example, the posterior mean, which is the point estimator of θ with a quadratic loss, is obtained with the choice $h(\theta) = \theta$. Analogous choices for $h(\theta)$ exist to obtain marginal posterior distributions, posterior median and credible intervals.

6.2.1 The quadrature technique

The quadrature technique is a numerical approximation method that consists of approximating integrals by finite sums. Consider a parameter vector of dimension $p = 2$, that is, $\theta = (\theta_1, \theta_2)$. In the simplest quadrature rule, the posterior marginal distribution of θ_1 and posterior expectation of any function $h(\theta_1)$ can be approximated, respectively, by

$$p(\theta_1 \mid y) = \int_{\Theta_2} p(\theta \mid y) \, d\theta_2 = \frac{\int_{\Theta_2} l(\theta; y) \pi(\theta) \, d\theta_2}{\int_{\Theta} l(\theta; y) \pi(\theta) \, d\theta}$$

$$\approx \frac{\int_{l_2}^{u_2} l(\theta; y) \pi(\theta) \, d\theta_2}{\int_{l_1}^{u_1} \int_{l_2}^{u_2} l(\theta; y) \pi(\theta) \, d\theta_2 \, d\theta_1} \approx \frac{\sum_{j_2=0}^{L_2} l(\theta_1, t_{j_2}; y) \pi(\theta_1, t_{j_2}) \Delta_{j_2}}{\sum_{j_1=0}^{L_1} \sum_{j_2=0}^{L_2} l(t_{j_1}, t_{j_2}; y) \pi(t_{j_1}, t_{j_2}) \Delta_{j_1} \Delta_{j_2}}$$

$$(6.13)$$

and

$$E(h(\theta_1) \mid y) = \int_{\Theta_1} h(\theta_1) p(\theta_1 \mid y) d\theta_1$$

$$(6.14)$$

$$\approx \int_{l_1}^{u_1} h(\theta_1) p(\theta_1 \mid y) d\theta_1 \approx \sum_{j_1=0}^{L_1} h(t_{j_1}) p(t_{j_1} \mid y) \Delta_{j_1},$$

where $\Delta_{j_k} = t_{j_k} - t_{j_k-1}$ and $l_k = t_0 < t_1 < \cdots < t_{L_k} = u_k$, for $k = 1, 2$, and $(l_1, u_1) \times (l_2, u_2) \subset \mathbb{R}^2$. The integration intervals (l_1, u_1) and (l_2, u_2) have to be specified in the case of unbounded parameter spaces. Appropriate specifications of these intervals are such that $p(\theta \mid y)$ has negligible probability mass in the residual parameter space $\Theta - (l_1, u_1) \times (l_2, u_2)$.

The greater L_1 and L_2 are, the more precise are the results provided by the method. This means that the numerical error may be as small as established by the user by increasing L_1 or L_2. Other summary statistics as the variance, the quantiles and the credible intervals may also be approximated by the quadrature technique. There are, in the literature, many other proposals of quadrature approaches supporting statistical inference. Some examples are the Gaussian, Newton-Cotes and Gauss-Hermite quadrature techniques. Detailed descriptions of these approaches and other numerical integration procedures may be found in Migon et al. (2014).

The quadrature technique is easily extended to dimensions larger than 2 by repeated application of the same rule used above to all dimensions. It is a powerful inference tool when θ is a vector with low dimension. Unfortunately, this procedure is practicable only for the single-digit dimension Θ, because the number of required numeric operations increases exponentially with respect to dimension p. This practical constraint of the quadrature method is an example of what is usually referred to in the literature as the curse of dimensionality. This issue is directly caused by the exponentially large number of sub-regions that must be considered and can be mitigated by careful elimination of all sub-regions that amount to negligible probability. This task requires further calculations and must be exercised with care. Other methods to estimate models with high-dimensional posterior distributions are briefly presented in the next section.

6.2.2 Markov Chain Monte Carlo simulation methods

This section presents a summarised description of some Markov Chain Monte Carlo (MCMC) sampling methods.

A Markov chain is a sequence of random variables $x_1, x_2, \ldots$ with the property that, for all $i \in \mathbb{N}$, the probability distribution of x_i, given all random variables $x_1, \ldots, x_{i-1}$, depends only on the random variable in position $i - 1$, that is, $p(x_i \mid x_{i-1}, \ldots, x_1) = p(x_i \mid x_{i-1})$. The MCMC sampling methods rely on a Markov chain with specific limiting distribution. After a sufficiently large number of iterations, a realisation of this chain will approximately contain samples of the target probability distribution.

Bayesian inference can be approximately performed by considering this target distribution to be the posterior $p(\theta \mid y)$, simulating values from a Markov chain with this posterior as the chain limiting distribution, and then computing summaries of the simulated values of θ. R is an example of software that provides many functions that simulate samples of a large number of known probability distributions. These functions are essential to the application of the sampling methods described next.

The Metropolis-Hastings algorithm (M-H) and Gibbs sampler are powerful examples of MCMC methods to simulate from high-dimensional distributions. The M-H was introduced by Metropolis et al. (1953) and Hastings (1970). The Gibbs sampler was proposed by Geman and Geman (1984) and popularised by a study of its properties by Gelfand and Smith (1990). In the M-H method, samples are sequentially generated from an arbitrary auxiliary Markov transition kernel (that can be easily sampled using the R software, for example) and each of these samples is accepted or rejected based on a given acceptance probability based on the kernel and the posterior density. If it is accepted, it becomes a new element of the chain; otherwise, the previous sampled value is repeated as a new element of the chain.

Denote by $q(\cdot \mid \theta)$ an arbitrary transition kernel that is easy to sample from and denote by s the iteration index. The M-H algorithm to generate a sample from $p(\theta \mid y)$ can be implemented with the steps below:

1. Set $s = 1$ and an arbitrary initial value $\theta^{(0)} = (\theta_1^{(0)}, \ldots, \theta_p^{(0)})$.

2. Sample θ' from $q(\cdot \mid \theta^{(s-1)})$.

3. Compute the acceptance probability

$$\alpha(\theta^{(s-1)}, \theta') = \min\left\{1, \frac{p(\theta' \mid y)}{p(\theta^{(s-1)} \mid y)} \frac{q(\theta^{(s-1)} \mid \theta')}{q(\theta' \mid \theta^{(s-1)})}\right\}.$$

4. Generate $u \sim U(0,1)$. If $u \leq \alpha(\theta^{(s-1)}, \theta')$, accept $\theta^{(s)} = \theta'$; otherwise, $\theta^{(s)} = \theta^{(s-1)}$.

5. Set $s = s + 1$ and return to step 2 until convergence is reached.

The Gibbs sampler is a MCMC method that produces a Markov chain with stationary distribution given by a target distribution and a transition kernel based on full conditional distributions of the target. Assume the full conditional distributions $p(\theta_i \mid \theta_1, \ldots, \theta_{i-1}, \theta_{i+1}, \ldots, \theta_p, y)$, for $i = 1, \ldots, p$, are known and can be sampled from. The main steps of the algorithm are as follows:

1. Set $s = 1$ and an arbitrary initial value $\theta^{(0)} = (\theta_1^{(0)}, \ldots, \theta_p^{(0)})$.

2. Generate $\theta^{(s)} = (\theta_1^{(s)}, \ldots, \theta_p^{(s)})$ successively as indicated below:

$$\theta_1^{(s)} \sim p(\theta_1 \mid \theta_1^{(s-1)}, \ldots, \theta_p^{(s-1)}, y),$$

$$\theta_2^{(s)} \sim p(\theta_2 \mid \theta_1^{(s)}, \theta_3^{(s-1)}, \ldots, \theta_p^{(s-1)}, y),$$

$$\vdots$$

$$\theta_p^{(s)} \sim p(\theta_p \mid \theta_1^{(s)}, \ldots, \theta_{p-1}^{(s)}, y).$$

3. Set $s = s + 1$ and return to step 2 until convergence is reached.

A chain constructed via M-H or Gibbs sampler is a Markov chain because the sample in iteration s only depends on the previous iteration $s - 1$. When convergence is reached, each $\theta^{(s)}$ is a sample of the posterior distribution $p(\theta \mid y)$. Descriptions and detailed theoretical formulation of these methods can be found in Gamerman and Lopes (2006) and Liu (2004). These references also present methods to increase the performance of these algorithms.

These MCMC methods may present undesirable practical inefficiency regarding the number of iterations s required to achieve approximate convergence of the chain to $p(\theta \mid y)$. The necessary number of iterations may be too large, making the method impracticable in some situations. The chain is expected to have good mixing properties, loosely meaning low chain autocorrelation. With poor mixing, the chain will exhibit high autocorrelation, producing samples potentially concentrated in specific parts of the parameter space or requiring an unfeasibly large number of iterations. In this case, the resulting chain will not be computationally viable to produce a good representation of $p(\theta \mid y)$.

The convergence of a chain may be verified through visual inspection of the series formed by the chain values. It is usually recommended to generate more than one chain, with different arbitrary initial values to reduce the possibility of convergence to local, minor modes in regions of negligible probability. Convergence is considered attained when the sample values of all the chains become equally dispersed around the same parameter region along the chain values. The period in which the chain has not yet reached convergence is called the *warm-up* period, also known as the *burn-in* period. These initial values are not a representative sample of the target distribution, and must be discarded.

As mentioned above, the presence of autocorrelation in the chain is undesired. This issue can be circumvented by extracting the values of the chain according to a fixed iteration length. Then, consider the new chain formed only by these extracted values. This distance should be large enough such that no significant autocorrelation is found in the new chain. This procedure is known as *thinning*. Because the new chain is a subset of the original chain, it is necessary to increase the number of MCMC iterations so that the posterior sample will have the desired length for inference. The *thinning* procedure contributes to a resulting chain with a higher number of nearly independent values, that is known as the *effective sample size*. The ideal situation is to have all the samples of the chain as independent samples. Another useful measure for convergence diagnostics is the *potential scale reduction*, which indicates whether a sample bigger than the current one may improve the inference or not. The *effective sample size* and the *potential scale reduction* are detailed in Gamerman and Lopes (2006), together with many other techniques to evaluate the convergence of MCMC methods. Some of these techniques are applied to real data sets in later parts of this book. Graphical analyses are also presented.

Among the many methodological extensions that have been proposed during the last decades to improve the efficiency of MCMC methods, the Hamiltonian Monte Carlo (HMC) method (also called the hybrid Monte Carlo), introduced by Duane et al. (1987), has a fundamental role in the construction of the PandemicLP package presented in Chapter 15. An improved variant of the HMC method, called NUTS (No-U-Turn Sampling), proposed by Hoffman and Gelman (2014), can be applied via the open-source Stan software (mc-stan.org). Codes and other specialties related to the practical use of Stan will be discussed in Chapter 10.

The HMC theoretical formulation is beyond the scope of this book. In short, it samples from a smartly chosen proposal density $q(\cdot)$ to be used in a Metropolis-Hastings setting. For this purpose, the model is augmented with potential and kinetic energy components and an imaginary particle travels along the geometry of the posterior distribution while maintaining the sum of these augmented components constant, according to the Hamiltonian dynamic. The intuitive idea is to imitate the behaviour of a particle travelling along the surface of a multi-dimensional upside-down "bowl". Then, the proposal distribution is centered on this particle's position. The NUTS variant of the HMC method avoids the undesirable possibility of the imaginary particle returning to the starting point of the algorithm on the surface, which would reduce the sampling performance. The reader can find a detailed description of this methodology, and about how it is implemented in Stan, in Gelman et al. (2013).

Similar to Stan, other computer programs have also been constructed to facilitate model estimation via MCMC methods. Some other notable examples are WinBUGS (mrc-bsu.cam.ac.uk/bugs), OpenBUGS (www.openbugs.net), JAGS (mcmc-jags.sourceforge.net) and NIMBLE (r-nimble.org). All these software programs are powerful tools to estimate a large family of Bayesian models. It is relevant to mention that their use is sometimes restricted to models with continuous parametric spaces, that is, these programs may not be feasible to run in the presence of discrete or categorical parameters. All the pandemic models discussed in this book are restricted to continuous parametric spaces, which makes all of these software programs practicable to be applied.

Among a host of alternative methodologies to MCMC, the integrated nested Laplace approximation (INLA) introduced by Rue et al. (2009) is highlighted. This method may be implemented via the R-INLA package and is useful for the class of latent Gaussian models. Detailed documentation about this methodology and the R-INLA package is available in www.r-inla.org.

The main features of the computer software cited in this section are discussed in Chapter 10.

Bibliography

Bernardo, J. M. and Smith, A. F. M. (2000) *Bayesian Theory.* Wiley Series in Probability and Statistics. New York: John Wiley & Sons.

Birnbaum, A. (1962) On the foundations of statistical inference. *Journal of the American Statistical Association*, **57**, 269–306.

Dawid, A. P. (1991) Fisherian inference in likelihood and prequential frames of reference. *Journal of the Royal Statistical Society. Series B (Methodological)*, **53**, 79–109.

Duane, S., Kennedy, A., Pendleton, B. J. and Roweth, D. (1987) Hybrid Monte Carlo. *Physics Letters B*, **195**, 216–222.

Gamerman, D. and Lopes, H. F. (2006) *Markov Chain Monte Carlo Stochastic Simulation for Bayesian Inference.* New York: Chapman and Hall/CRC, 2nd edn.

Gelfand, A. and Smith, A. (1990) Sampling based approaches to calculating marginal densities. *Journal of the American Statistical Association*, **85**, 398–409.

Gelman, A., Carlin, J. B., Stern, H. S., Dunson, D. B., Vehtari, A. and Rubin, D. B. (2013) *Bayesian Data Analysis.* New York: Chapman and Hall/CRC, 3rd edn.

Gelman, A., Simpson, D. and Betancourt, M. (2017) The prior can often only be understood in the context of the likelihood. *Entropy*, **19**.

Geman, S. and Geman, D. (1984) Stochastic relaxation, Gibbs distributions and the Bayesian restoration of images. In *IEEE Transactions Pattern Analysis and Machine Intelligence*, vol. 6, 721–741.

Gneiting, T., Balabdaoui, F. and Raftery, A. E. (2007) Probabilistic forecasts, calibration and sharpness. *Journal of the Royal Statistical Society: Series B (Statistical Methodology)*, **69**, 243–268.

Hastings, W. (1970) Monte Carlo sampling using Markov chains and their applications. *Biometrika*, **57**, 97–109.

Hoffman, M. D. and Gelman, A. (2014) The No-U-Turn sampler: Adaptively setting path lengths in Hamiltonian Monte Carlo. *Journal of Machine Learning Research*, **15**, 1351–1381.

Liu, J. S. (2004) *Monte Carlo Strategies in Scientific Computing*. Springer Series in Statistics. New York: Springer-Verlag New York, 1st edn.

Metropolis, N., Rosenbluth, A., Teller, M. and Teller, E. (1953) Equations of state calculations by fast computing machines. *Journal of Chemistry and Physics*, **21**, 1087–1091.

Migon, H. S., Gamerman, D. and Louzada, F. (2014) *Statistical Inference: An Integrated Approach*. New York: Chapman and Hall/CRC, 2nd edn.

O'Hagan, A., Buck, C. E., Daneshkhah, A., Eiser, J. R., Garthwaite, P. H., Jenkinson, D. J., Oakley, J. E. and Rakow, T. (2006) *Uncertain Judgements: Eliciting Experts' Probabilities*. West Sussex: John Wiley & Sons.

Robert, C. (2007) *The Bayesian Choice: From Decision-Theoretic Foundations to Computational Implementation*. Springer Texts in Statistics. New York: Springer-Verlag, 2nd edn.

Rue, H., Martino, S. and Chopin, N. (2009) Approximate Bayesian inference for latent Gaussian models by using integrated nested Laplace approximations. *Journal of the Royal Statistical Society: Series B (Statistical Methodology)*, **71**, 319–392.

Schervish, M. J. (1995) *Theory of Statistics*. Springer Series in Statistics. New York: Springer-Verlag, 1st edn.

Part III

Further Modelling

7

Modelling misreported data

Leonardo S. Bastos

Fundação Oswaldo Cruz, Brazil

Luiz M. Carvalho

Fundação Getulio Vargas, Brazil

Marcelo F.C. Gomes

Fundação Oswaldo Cruz, Brazil

CONTENTS

During a pan/epidemic, data on cases and deaths must be opportune. That is, data collection and availability on official databases must be provided in a timely manner. To make data-driven actions, the faster the information is obtained, the better. However, the quality of information is inversely related to the speed the information is gathered. In this chapter, we describe some data quality issues like reporting delays and disease misclassification in the context of the COVID-19 pandemic and severe acute respiratory illness (SARI) surveillance in Brazil. Moreover, for each of these issues, a statistical model-based solution is presented to incorporate/propagate the uncertainty related

DOI: 10.1201/9781003148883-7

to the lack of data quality into the inference process, making the analysis more adequate. Although the examples are focused on those two specific contexts, they can be easily translated to other diseases with structured notification systems, being relevant for public health surveillance in general.

7.1   Issues with the reporting of epidemiological data

In any passive surveillance[1] system, be it related to outpatient cases such as influenza-like illness (ILI) and arboviruses surveillance, or to hospitalised cases such as severe acute respiratory illness (SARI) surveillance, there is an intrinsic time delay between the event of infection and its notification in the corresponding database. This is due to the fact that notified cases are only identified once symptomatic individuals are attended to at a health care unit. This delay will depend on characteristics of the disease itself, the typical incubation period which will define how long it takes for infected individuals to develop symptoms, as well as cultural and structural characteristics of the exposed population, which affects how long symptomatic individuals usually wait before seeking medical attention. In situations where the notification is not made automatically in the official database, there will also be a potential delay between the notification date (the date in which the case is identified at the healthcare unit) and the digitisation date (the date in which the notification sheet is typed into the digital database). It is clear then that notification delays have both a structural component that can be minimised by infrastructure investments, such as the hiring of dedicated staff for filling notification forms and inserting this information into the digital databases, migration from paper forms to electronic notification, good quality internet access at health care units, and so on (Lana et al., 2020); and an intrinsic component that is related to the patient itself which can be mitigated by ease-of-access to health care facilities and information campaigns to motivate early medical attention. To illustrate this process, we will describe the timeline of a hypothetical COVID-19 case captured by local surveillance.

Let's assume the following example: a person has just found out that he was recently in contact with someone infected by SARS-CoV-2. We shall assume this encounter occurred at time t_0. At time t_1, the patient presented the first symptoms: shortness of breath, fever, anosmia (no sense of smell) and ageusia (loss of taste). Then these symptoms get worst at time t_2 and he, the patient, sought medical assistance where he were tested and hospitalised. The test result was negative for SARS-CoV-2, but given their symptoms the doctor was quite convinced that it was a COVID-19 case. So she, the doctor, asked for another test. At time t_3, she received the test result and, as she expected, it

---

[1] Passive surveillance: that in which cases are reported as patients seek medical care.

was positive for SARS-CoV-2, that is, indeed a COVID-19 case. The attending health care professional should enter the case into a surveillance system, but when they received the result they were too busy to do it. Eventually they reported it at time t_4. The patient's condition got worse and he was transferred to the intensive care unit ICU at time t_5. Unfortunately, at time t_6, the patient passed away, which was then reported to the surveillance system at time t_7. The timeline is qualitatively represented in Figure 7.1.

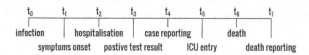

FIGURE 7.1: Qualitative timeline of events related to a hypothetical hospitalised COVID-19 case which required ICU and evolved to death.

The previous example illustrates a COVID-19 case from infection to death. Notice that it could be any infectious disease confirmed by a laboratory test. Recording the times $t_0, t_1, t_2, \ldots$ are essential for the surveillance team to learn about the dynamic of each epidemic. Surveillance systems gather information about several patients like the previous example in a structured data set. It is compulsory for diseases to be reported immediately as soon as a case is identified, for instance yellow fever, which is a mosquito-borne disease transmitted to humans by infected mosquitoes. There are some diseases for which it is compulsory hospitalised cases and/or deaths to be reported, like dengue fever, which is another mosquito-borne disease transmitted to humans very frequently in Latin America and Southeast Asia. And there are some diseases for which reporting is not compulsory, but for which there is a sentinel surveillance system that monitors cases, such as influenza-like illness in most countries in Latin America, the USA and in Europe. In this chapter, we will use COVID-19 as an example and assume that reporting of hospitalised cases and deaths is mandatory.

The time t_0 representing when the person was infected is usually unknown. A person may be able to describe a list of situations where they were exposed but it is nearly impossible to correctly say when the infection occurred. On the other hand, the time of onset symptoms, t_1, is more likely to be remembered by the patient and it is usually a time recorded in a surveillance system. Still there is some memory bias, since the patient may not be certain about when the first symptoms started (t_1). The time when the person seeks medical care is very clear, t_2, which would be the day when the person is evaluated by a health professional. This is usually called the notification date, and is also called the occurrence date. Once a case of a particular disease, COVID-19 in this case or at least a suspected case, is identified, it should be reported to the surveillance team. This notification may occur at the same time the person seeks assistance, but it may also be reported later on at time t_4, called the digitisation date in the case of digital databases. The delays given by the differences $t_4 - t_1$ and $t_4 - t_2$ are very important. The first delay represents the time until the surveillance team is aware of an infectious disease case. The longer it takes, the more people may be infected from this person. Hence, if the surveillance team is aware of this delay, then it may be able to warn the health units about an outbreak of an infectious disease occurring in a particular area and act in order to mitigate the disease spreading. The second delay, $t_4 - t_2$, is the time between a case being identified until it is reported in a surveillance system, reflecting the capacity of the health unit to report cases. For instance, suppose a health unit is in an isolated area without a computer. In this case, a notification is filed in a paper-based form and eventually it will be recorded in a surveillance system.

Naturally, the longer the notification/digitisation delay, the more inadequate the current case count is for alert systems and situation analysis based on daily or weekly incidence. Not only that, since current incidence will be underestimated, it also impacts the adequacy of this data as input for mathematical and statistical forecasting models. This is where the modelling of reporting delays comes into play, providing *nowcasting* estimates. That is, estimates for what happened up-to-now, in contrast to forecasting which estimates what will come next. Based on the distribution of these delays, the surveillance team may be able to predict the actual number of cases that occurred but have not been notified yet, plan public response in light of the current situation, as well as devise strategies in order to reduce these delays when they become overwhelming. Nowcasting models are also described as a backfill problem in the literature. The description of this process and application examples are described in Section 7.2.

Diseases like COVID-19 also require a biological test to confirm the infection, so on top of reporting delay, there is an extra time associated with testing. It could be a rapid test or a test run in a laboratory, and after knowing the result, the patient record in the surveillance system must be updated. So the surveillance team have to be careful while analysing counts of a disease that can only be confirmed after a biological test. The counts are affected by delay and even under notification since the information, namely here the test result, for some patients might not be updated in the surveillance system.

By definition, public health surveillance strategies based on syndromic definitions will potentially capture cases from multiple pathogens, not only cases related to a single one. In the case of SARI, several respiratory viruses are associated with those manifestations such as Influenza A and B, respiratory syncytial virus (RSV), adenovirus, the coronavirus family, including SARS-CoV-2, and many others. Therefore, SARI cases can be classified as a suspect case of infection by any of those viruses. In this scenario, sample collection for laboratory confirmation of the associated pathogen is fundamental for assessing incidence and prevalence of specific diseases, such as COVID-19. On the other hand, surveillance systems based on confirmed cases alone will not mix multiple pathogens in the same database, but case notification will only happen if and when positive laboratory results are obtained.

In the example given here, at time t_2 the patient was tested and the result of this test was reported to their doctor at time t_3, defined as the laboratory delay. And it is a structural delay since it depends on collecting, transporting and analysing a sample from the patient added to the time until the doctor and the patient get back the result. It also depends on the test itself, since there are both rapid tests and tests that need a proper laboratory infrastructure. Also the surveillance team is informed about the laboratory results of a patient with yet another possible delay. It could be automatically informed as soon as the result is known, but it could also be informed only after the patient and their medical doctor know the result, at which time the health professional should update the patient record in the surveillance system.

Notice that at any moment the sample may be damaged leading to a misclassified result; for example, it could be poorly handled by a technician without experience, or during the transport the sample might not be properly stored at the appropriate temperature, etc. Hence a person may get a false negative result due to any of these reasons and their combination. Even if the sample is not damaged, most laboratory tests are not perfect. They have non-null probabilities for false results. The time between exposure, symptoms onset, and sample collection also affect the ability to identify the associated pathogen by any given test, even if proper care regarding sample collection and handling were in place. And different tests have different optimal time windows (La Marca et al., 2020). The probability of a positive result when the patient is really infected is called sensitivity, whereas the probability of a negative result given the patient is not infected is called specificity. In Section 7.3, we present some statistical models to incorporate external information regarding imperfect tests in order to correct infection incidence and prevalence.

7.2 Modelling reporting delays

As illustrated in the previous section and also in Chapter 2, reporting delay is a well-known issue in infectious disease surveillance. The timeliness of case reporting is key for situation analysis based on surveillance data (Centers for Disease Control, 1988; World Health Organization, 2006; Lana et al., 2020). Although efforts should be made to have case reporting be as timely as possible, in practice it is always expected to have at least some level of delay, especially so for passive surveillance. In order to estimate current or recent cases taking into account the reporting delays described in the previous section, it is fundamental that the notification database store all the relevant dates related to the desired count. For suspected cases or general syndromic cases, the minimum information necessary are the symptoms onset and notification date. For simplicity, in this section we will assume that digitisation and notification dates are the same. When this is not the case, digitisation date is of the utmost relevance, since this is the actual date at which each case will be available in the database for situation analysis. Nonetheless, the presence of notification date is important for public health authorities in assessing to what extent the observed digitisation delay is due to administrative/structural issues, by evaluating the delay between notification and digitisation date. Since our focus here is to describe nowcasting techniques aimed at case counts for situation analysis and/or to be used as input for forecasting models, the only thing that matters is how long it takes for the information to be available in the database, not the intermediate steps themselves.

For confirmed cases, along with symptoms onset date, one would need access to not only the date of the test result but, more importantly, the date on which the result was inserted into the database for each notified case; call it the test result digitisation date. In the same way, for deaths one would need the date of death and the death digitisation date. For simplicity, we will discuss only the nowcast of suspected cases but, as long as those dates are available, the methods can easily be translated to nowcasting of confirmed cases or deaths.

To illustrate the limited information available by the end of each epidemiological week in a real setting, Figure 7.2 shows the time series of SARI data from Brazil during the 2020 season, consolidated at different epidemiological weeks: 15, 25, 35, 45, and 53, the last epidemiological week of 2020. It is clear that the incompleteness of recent weeks due to notification delay not only affects the magnitude, but also the current trend.

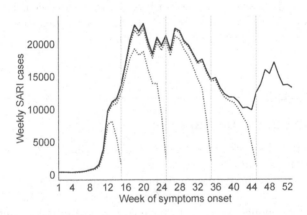

FIGURE 7.2: Weekly cases of severe acute respiratory illness (SARI) notified in Brazil during the 2020 season, by symptoms' onset week. Comparison between the weekly cases digitised up to the last epidemiological week, 53 (solid black line), and what had been digitised up to the end of epidemiological weeks 15, 25, 35, and 45 (dotted grey lines). Vertical lines indicate each of those weeks for reference.

7.2.1 Weekly cases nowcast

In this chapter we will assume that the quantity of interest is the weekly number of new cases, y_t. Nonetheless, the methodology can transparently be applied to any scale of interest, be it daily, weekly, or monthly. The process of estimating current or recent cases is defined as a nowcasting exercise, a nomenclature based on the term forecasting. While forecast is the exercise of estimating what is expected for the coming weeks, nowcasting tries to evaluate what has already happened but has not been reported yet.

Since cases can be inserted into the database retrospectively, at each following week $t + 1$, $t + 2$, ..., $t + n$, cases corresponding to week t can still be registered. Let us define $y_{t,d}$ the number of new cases that occurred at week t, based on symptoms onset, but inserted into the database d weeks after week t, based on digitising date. Therefore, the total number of notified cases from week t can be written as

$$y_t = \sum_{d=0}^{d_m} y_{t,d}, \tag{7.1}$$

where d_m is the maximum delay between symptom onset and digitisation. This can be defined by the surveillance team beforehand, or extracted from the data. As it will be clear in the discussion ahead, the higher d_m is, the more computationally expensive the nowcasting will become. Exploratory analysis of historical data can be used to define a cut-off $d_m > 0$ for which $y_{t,d_m} \approx 0$ or $y_{t,d_m} \ll y_t$ for all t as well.

The problem is that, for the most recent complete week T, the database will only have $y_{T,0}$. For the previous week, $T-1$, it will only have $y_{T-1,0}$ and $y_{T-1,1}$, and so on until week $T - d_m$, for which it will have the complete information $y_{T-d_m,0}$, $y_{T-d_m,1}$, ..., y_{T-d_m,d_m}. Therefore, the process of nowcasting can be described as the process of estimating the not-yet-available data $y_{T,1}$, $y_{T,2}$, ..., y_{T,d_m}, $y_{T-1,2}$, ..., y_{T-1,d_m}, ..., y_{T-d_m+1,d_m}. By representing the information in a matrix where lines are the weeks increasing downwards, and columns are the delay increasing from left to right, the missing cells would form a nice triangle, known as the runoff triangle (Mack, 1993):

$$
\begin{array}{c}
 \\
T - d_m \\
T - d_m + 1 \\
T - d_m + 2 \\
\vdots \\
T - 1 \\
T
\end{array}
\begin{pmatrix}
0 & 1 & \cdots & d_m - 1 & d_m \\
y_{T-d_m,0} & y_{T-d_m,1} & \cdots & y_{T-d_m,d_m-1} & y_{T-d_m,d_m} \\
y_{T-d_m+1,0} & y_{T-d_m+1,1} & \cdots & y_{T-d_m+1,d_m-1} & \text{NA} \\
y_{T-d_m+2,0} & y_{T-d_m+2,1} & \cdots & \text{NA} & \text{NA} \\
\vdots & \vdots & \vdots & \vdots & \vdots \\
y_{T-1,0} & y_{T-1,1} & \cdots & \text{NA} & \text{NA} \\
y_{T,0} & \text{NA} & \cdots & \text{NA} & \text{NA}
\end{pmatrix}
\tag{7.2}
$$

There are several ways to tackle this problem, and most approaches can be broadly grouped in two approaches: ones that are hierarchical in nature, modelling $y_{t,d}$ conditional on y_t, where the latter is given by a Poisson or Negative Binomial distribution and the former is a multinomial process with probability vector of size d_m (Noufaily et al., 2016; Höhle and an der Heiden, 2014); and others that focus on the distribution of the random variables $y_{t,d}$ themselves (Bastos et al., 2019; Barbosa and Struchiner, 2002), also known as the chain-ladder technique from actuarial sciences (Mack, 1993; Renshaw and Verrall, 1998).

In this chapter we will focus on the method described in Bastos et al. (2019), a Bayesian approach to estimate $y_{t,d}$ which was successfully applied to SARI and Dengue surveillance in Brazil, generating a publicly available weekly nowcast reported by InfoGripe (`http://info.gripe.fiocruz.br`) and Info-Dengue (`http://info.dengue.mat.br`) (Codeço et al., 2018) systems. The implementation of this method for SARI surveillance within InfoGripe was able to provide an early warning for the impact of COVID-19 cases in Brazil even before laboratory confirmation of SARS-CoV-2 predominance among SARI cases (Bastos et al., 2020).

We assume $y_{t,d}$ follows a Negative Binomial distribution with mean $\mu_{t,d}$ and scale parameter ϕ

$$y_{t,d} \sim \text{NegBin}(\mu_{t,d}, \phi), \quad \mu_{t,d} > 0, \ \phi > 0. \tag{7.3}$$

The adopted parameterisation provides $\text{Var}[y_{t,d}] = \mu_{t,d}(1 + \mu_{t,d}/\phi)$. The logarithm of the mean of $y_{t,d}$ can be decomposed in a way that takes into account random effects associated with a mean temporal evolution α_t, the delay effect β_d, interaction effects between the two $\gamma_{t,d}$ to accommodate structural changes affecting digitisation timeliness, as well as other relevant temporal covariates $\mathbf{X}_{t,d}$ such as weather data, for example, with its corresponding parameters vector $\vec{\delta}$. Therefore, it can be described as

$$\log(\mu_{t,d}) = \mu + \alpha_t + \beta_d + \gamma_{t,d} + \mathbf{X}'_{t,d}\vec{\delta}, \tag{7.4}$$

where μ is the logarithm of the overall mean number of weekly cases.

The temporal and delay effects can be modelled as a simple first-order random walk

$$\alpha_t \sim N(\alpha_{t-1}, \sigma_\alpha^2), \quad t = 2, 3, \dots, T, \tag{7.5}$$

$$\beta_d \sim N(\beta_{d-1}, \sigma_\beta^2), \quad d = 1, 2, \dots, d_m. \tag{7.6}$$

The interaction between time and delay is also modelled as a first-order random walk

$$\gamma_{t,d} \sim N(\gamma_{t-1,d}, \sigma_\gamma^2). \tag{7.7}$$

Since timeliness can be affected by several factors over time, it is important to incorporate this interaction effect. For example, during peaks of hospital capacities, it is possible to have an increase in the notification delay if there is no dedicated staff for filling out notification sheets, since patient care is prioritised by medical staff. Conversely, if local authorities hire dedicated staff in anticipation or as a response to an outbreak, delay can be reduced. Note that other factors can be easily integrated into the model described in Eq 7.4, including spatial effects (Bastos et al., 2019).

Given the described model, the posterior distribution of the parameters, $\Theta = (\mu, \{\alpha_t\}, \{\beta_d\}, \{\gamma_{t,d}\}, \sigma_\alpha^2, \sigma_\beta^2, \sigma_\gamma^2, \phi)$ given the set of notified data $\mathbf{y} = \{y_{t,d}\}$ can be expressed by

$$p(\Theta|\mathbf{y}) \propto \pi(\Theta) \prod_{t=1}^{T} \prod_{d=0}^{d_m} p(y_{t,d}|\Theta), \qquad (7.8)$$

with $p(y_{t,d}|\Theta)$ the Negative Binomial defined in Equation 7.3, and $\pi(\Theta)$ the joint prior distribution given by the random effects distributions and the prior distributions of their corresponding scale parameters.

Samples from the posterior distribution $p(\Theta|\mathbf{y})$ can be obtained by integrated nested Laplace approximation (Rue et al., 2009; Martins et al., 2013) (INLA), for which there is a publicly available package for R at `https://www.r-inla.org`. Although this process could also be implemented using classical Monte Carlo Markov Chain (MCMC) models (Gamerman and Lopes, 2006), INLA provides a significantly accelerated performance, which is a must for its usefulness in situation rooms during ongoing surveillance, especially when dealing with analysis from multiple locations at once such as those from state or national health secretariats. This characteristic was paramount for the feasibility of weekly nowcast of arboviruses and SARI notifications at the municipal, state and national levels implemented by Info-Dengue (Codeço et al., 2018) (`http://info.dengue.mat.br`) and InfoGripe (`http://info.gripe.fiocruz.br`) in Brazil, as illustrated in Section 7.2.2.

With the samples from the posterior for the parameters obtained from Eq 7.8, for example by means of `inla.posterior.sample()` function, it is a simple question of implementing a Monte Carlo process to generate a sample of the missing values $\{y_{t,d}\}$ from the matrix illustrated in Equation 7.2. From those, we obtain the estimated marginals $\{\hat{y}_t| \ t = T-d_m+1, T-d_m+2, \ldots, T\}$, which is the actual nowcast of weekly cases. For each generated sample of the posterior distribution of the parameters, we have an estimated trajectory for the weekly cases. Therefore, this process allows for point estimates and credible intervals for the weekly cases. This can then be used to propagate the uncertainty to predictive or forecasting models discussed in other chapters in this book.

Open-source codes for implementing this model can be found at two main repositories maintained by the authors of this chapter: `https://github.com/Opportunity-Estimator-EpiSurveillance/leos.opportunity.estimator` and `https://github.com/lsbastos/Delay`.

7.2.2   Illustration: SARI notifications in Brazil provided by InfoGripe

The InfoGripe platform (`http://info.gripe.fiocruz.br`) was developed as a joint effort by Brazilian researchers from Fiocruz[2] and FGV[3] with the Influenza Technical Group of the Health Surveillance Secretariat (*GT-Influenza, Secretaria de Vigilância em Saúde*) from the Ministry of Health. It is used as an analytical tool for the national surveillance of SARI in Brazil (Ministério da Saúde et al., 2018), providing situation assessment by means of alert levels based on current incidence, typical seasonal profile by state, epidemiological data such as age and sex stratification, weekly cases by respiratory virus, and so on.

The current SARI surveillance scope in Brazil is based on notification from every health care unit with hospital beds (Ministério da Saúde et al., 2019), following a syndromic case definition inspired by the World Health Organization guidelines for SARI surveillance (World Health Organization, 2013). It started in 2009 as a response to the 2009 H1N1pdm09 Influenza pandemic and, in 2012, laboratory testing protocol was extended to include other respiratory viruses of interest (Ministério da Saúde, SINAN, 2012). In Figure 7.3 we show the point estimates for the weekly cases, $\hat{y}_t$ (dashed lines), provided by InfoGripe using the method described in the previous section. We compared it to the consolidated time series, y_t (solid lines), and the weekly counts for each week available by the end of the corresponding week, $y_{t,0}$ (dotted lines), for the aggregated data for Brazil as well as state counts for selected states.

In the face of the COVID-19 pandemic in 2020, it was only natural to use this surveillance system to monitor the corresponding hospitalisations and to incorporate SARS-CoV-2 testing for differential diagnosis (Ministério da Saúde and Secretaria de Vigilância em Saúde, 2020). Even before SARS-CoV-2 testing for notified SARI cases was widely implemented in Brazil, the nowcast model, along with the complementary epidemiological information provided by InfoGripe, were able to provide early warning of the impact of COVID-19 in terms of SARI cases in Brazil (Bastos et al., 2020), illustrating the usefulness of this methodology for action planning and response. Since the modelling approach provides estimated trajectories, it can also be used to assess the trend itself, which can be even more relevant than the actual count estimates. Early detection of a trend indicating an increase in the number of weekly cases, even if the actual number of cases is relatively low, allows for rolling out of mitigation strategies to prevent the collapse of medical resources and high disease burden. With that in mind, the InfoGripe platform was extended to provide short- and long-term trend based on a linear model applied as a rolling window to the last 3 and 6 consecutive weeks, respectively. This information was provided on a weekly basis in the form of situation reports

---

[2] *Fundação Oswaldo Cruz*, Fiocruz: `https://portal.fiocruz.br/`
[3] *Fundação Getúlio Vargas*, FGV: `https://portal.fgv.br/`

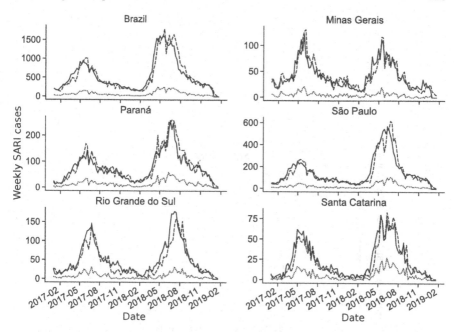

FIGURE 7.3: Weekly cases of severe acute respiratory illness (SARI) notified in Brazil during the 2017 and 2018 seasons, by symptoms' onset week. Comparison between the consolidated weekly cases, y_t (solid line), the point estimated provided by the end of each corresponding week, $\hat{y}_t$ using the described nowcasting model (dashed line), and the number of cases digitised up to the end of each epidemiological week, $y_{t,0}$ (dotted line), for the whole country (top left panel), the states of Minas Gerais (top right panel), Paraná (centre left), São Paulo (centre right), Rio Grande do Sul (bottom left), and Santa Catarina (bottom right). The first two are within the Southeastern region, while the remaining three are in the South region of Brazil.

forwarded to the Health Surveillance Secretariat at the Brazilian Ministry of Health (SVS/MS), and State Health Secretariats subscribed to InfoGripe's mailing list, as well was deposited in a public online repository (`http://bit.ly/mave-infogripe`). Figure 7.4 shows a snapshot taken from the bulletin published by the end of epidemiological week 37, as an illustration of real case application on the field. It provides the nowcast of the weekly incidence in the municipality of Manaus, capital of Amazonas state, along with the likelihood of increasing/decreasing trend in the short and long term up to weeks 36 and 37, and up to weeks 32 to 37, respectively.

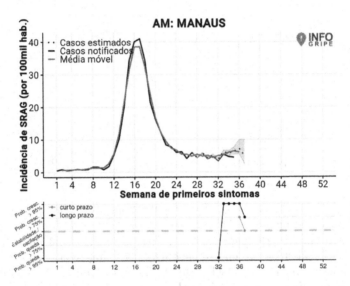

FIGURE 7.4: Snapshot from InfoGripe's Bulletin from epidemiological week 37 of 2020, available at `https://gitlab.procc.fiocruz.br/mave/repo/` `-/blob/master/Boletins%20do%20InfoGripe/boletins_anteriores/` `Boletim_InfoGripe_SE202037_sem_filtro_febre.pdf`. The upper plot shows the weekly incidence per 100k inhabitants of Manaus, capital of the Amazonas state, by symptom onset, with notified cases (*Casos notificados*, solid black line), estimated incidence (*Casos estimados*, dotted black line) with 90%CI, and centred 3-week rolling average of the point estimate (*Média móvel*, solid grey line). The bottom plot presents the trends in terms of likelihood of increase (*Prob. cresc.*) or decrease (*Prob. queda*) for the short term (*curto prazo*, grey line) and long term (*longo prazo*, black line).

7.3 Prevalence estimation from imperfect tests

An important question during an epidemic is how many individuals have been exposed to the disease at time t, i.e., the (true) number of cumulative cases, Y_t^{true}. Let M denote the population size. Assuming permanent and complete immunity, the proportion $\theta_t = Y_t^{\text{true}}/M$, called the **prevalence**, is a key epidemiological quantity because it measures the overall immunity of the population and whether the threshold for collective (herd) immunity has been reached.

However, factors such as reporting delays, underreporting and a large fraction of asymptomatic individuals might make it difficult to ascertain Y_t^{true} accurately. In such a scenario, serological and other diagnostic-based surveys provide a valuable source of additional data that can be used to uncover the underlying pattern of immunity to the disease in a population. The chief idea is to randomly sample individuals from the target population and, using a diagnostic test, measure their levels of antibodies against the causative agent of the disease in question. Assuming for simplicity that all tests are performed at time t, the number x of individuals that test positive out of n provides information about θ_t. Once θ_t has been estimated, we can project the number of exposed individuals $\hat{I}_t = M\hat{\theta}_t$. Unfortunately, diagnostic tests are rarely perfect measurement instruments; one might obtain false positives or false negatives (see Section 7.3.1). One thus needs a statistical model in order to account for this misclassification error.

This section is concerned with providing a statistically principled, model-based treatment of prevalence estimation, having the explicit goal of estimating Y_t with the appropriate level of uncertainty. The theory laid out here is general and encompasses any imperfect test, not just serological diagnostic tests, although the latter provide the main motivation for the techniques discussed.

7.3.1 Preliminaries: Imperfect classifiers

Before we discuss prevalence estimation, it is useful to establish the basic concepts and notation pertaining to (imperfect) binary classifiers. Let $R_i \in \{0, 1\}$ be the random variable which records whether the i-th individual in the sample has a given condition which we are interested in detecting. Also, let $T_i \in \{0, 1\}$ be the random variable which records whether the same individual will produce a positive test ($T_i = 1$). An imperfect test can be evaluated in the presence of a **gold standard**. The gold standard is a measurement device with the highest accepted accuracy and which can be taken to represent the ground truth about R_i. In the context of infectious diseases, a gold standard can be a very precise molecular test that detects the presence of the pathogen's genetic material, polymerase chain reaction (PCR) for instance.

If we measure J individuals with both the test under evaluation and the gold standard, we can tally the numbers of true positives/negatives (TP and TN, respectively) and false positives/negatives (FP and FN) such that TP + FP + TN + FN = J and produce a two-by-two table as the one in Table 7.1.

TABLE 7.1: Hypothetical counts from a (binary) diagnostic test against a gold standard. Here, a value of 1 represents a positive result, and a value of 0 represents a negative result.

		Test under evaluation	
		0	1
Gold Standard	0	TN	FP
	1	FN	TP

The counts from Table 7.1 can in turn be used to estimate key probabilistic quantities such as the sensitivity and specificity of the test. These quantities, their definitions and how to estimate them from a gold standard essay are shown in Table 7.2. As we shall see in the next sections, the sensitivity (δ) and specificity (γ) of a test are the key quantities to consider when estimating the prevalence. The positive and negative predictive values (PPV and NPV) are useful at an individual level because they measure the probability that one has (does not have) the condition being tested conditional on a positive (negative) test outcome. An important aspect of diagnostic tests is that their accuracy for measuring the status of any particular individual depends on the underlying prevalence of the condition being tested. In other words, the PPV and NPV of a test depend on the prevalence θ_t. This means that the inferential value of the results to which any one individual taking the test will strongly depend on the underlying prevalence. If the prevalence is high, the results are accurate at an individual level, but if the disease/condition is rare, the test has little value to inform whether one has the condition in question. This will hopefully become clear in the next section. For a review of the clinical applications of predictive values, sensitivity and specificity, see Parikh et al. (2008).

7.3.2 Prevalence from a single imperfect test

Having established the main accuracy measures of a (binary) test, we can now move on to how to estimate the prevalence θ_t from test data alone, without the gold standard. Estimation of prevalence from imperfect tests is a long-standing and well-studied problem in the Medical Statistics literature (Rogan and Gladen, 1978; Greenland, 1996; Diggle, 2011) and, as we shall see, follows naturally from basic probability manipulations.

TABLE 7.2: Key probabilistic quantities in imperfect classification. We show the probabilistic definition of each quantity, as well as how to estimate it from the counts in Table 7.1.

Quantity	Definition	Estimate
Sensitivity	$\delta := \mathrm{P}\left(T = 1 \mid R = 1\right)$	$\frac{\mathrm{TP}}{\mathrm{TP+FN}}$
Specificity	$\gamma := \mathrm{P}\left(T = 0 \mid R = 0\right)$	$\frac{\mathrm{TN}}{\mathrm{TN+FP}}$
Positive predictive value (PPV)	$\zeta := \mathrm{P}\left(R = 1 \mid T = 1\right)$	$\frac{\mathrm{TP}}{\mathrm{TP+FP}}$
Negative predictive value (PPV)	$\xi := \mathrm{P}\left(R = 0 \mid T = 0\right)$	$\frac{\mathrm{TN}}{\mathrm{TN+FN}}$

For concreteness, we will focus on the case where one surveys a sample of n individuals with a test that detects antibodies against a given pathogen. In this context we thus define R_i to be the random variable which records whether the i-th individual, once exposed, will produce a measurable immune response in the form of antibodies. Recalling the notation in the previous section, we say that the prevalence, $\theta_t := \mathrm{P}(R = 1)$, is the probability that an individual has antibodies and is the main quantity of interest (q.o.i). We can compute $p := \mathrm{P}(T_i = 1)$, sometimes called the **apparent prevalence**:

$$p = \mathrm{P}(T_i = 1 \mid R_i = 1)\,\mathrm{P}(R_i = 1) + \mathrm{P}(T_i = 1 \mid R_i = 0)\,\mathrm{P}(R_i = 0),$$
$$= \delta\theta_t + (1 - \theta_t)(1 - \gamma), \tag{7.9}$$

where δ and γ are the **sensitivity** and **specificity** of the test, respectively. This computation amounts to marginalising over the unobserved quantities in the model and thus can be seen as a Rao-Blackwellisation of the full model, i.e, $p = E[E[T_i|R_i]] = \sum_{r=0}^{1} \mathrm{P}(T_i = 1, R_i = r)$, where the first equality follows from the law of total expectation. The relevant (sufficient) statistic from this experiment is $x = \sum_{i=1}^{n} T_i$. Assuming the tests are conditionally independent given the latent status variables $\boldsymbol{R} = \{R_1, \ldots, R_n\}$, we conclude that x has a binomial distribution with n trials and success probability p. This assumption is reasonable when the sample is random and $n \ll M$. If sampling is not random or if n is large, then the dependency between individuals induced by the contagion process will not be negligible.

Frequentist analysis

A first stab at estimating θ_t can be made by correcting the naïve estimate $\hat{p} = x/n$, along with its usual normal approximation confidence interval, for the sensitivity and specificity of the test. A straightforward manipulation of (7.9) gives rise to the Rogan-Gladen estimator (Rogan and Gladen, 1978):

$$\hat{\theta}_t^{\mathrm{RG}} = \frac{\hat{p} - (1 - \gamma)}{\delta + \gamma - 1}. \tag{7.10}$$

Notice that if $\hat{p}$ is less than the false positive rate (FPR) of the test, $1 - \gamma$, the Rogan-Gladen estimator will yield a meaningless negative estimate. One might also wish to provide interval estimates for θ_t. For a confidence level $0 < \alpha < 1$, if one constructs a $100 \times \alpha\%$, confidence interval $(\hat{p}_l, \hat{p}_u)$ for p, an interval $(\hat{\theta}_l, \hat{\theta}_u)$ can be obtained for θ (Diggle, 2011):

$$\hat{\theta}_l = \max\left\{0, \frac{\hat{p}_l - (1 - \gamma)}{\delta + \gamma - 1}\right\}, \tag{7.11}$$

$$\hat{\theta}_u = \min\left\{1, \frac{\hat{p}_u - (1 - \gamma)}{\delta + \gamma - 1}\right\}, \tag{7.12}$$

under the reasonable assumption that $\delta + \gamma > 1$. This approach however ignores uncertainty about γ and δ and thus tends to produce overly-confident estimates (Izbicki et al., 2020). Whilst this can be remedied with bootstrap techniques (see Cai et al. (2020)), a Bayesian approach is able to seamlessly incorporate all the sources of information in order to produce sensible estimates (Flor et al., 2020).

Bayesian analysis

As demonstrated by Greenland (2009) and Gelman and Carpenter (2020) amongst many others, taking a Bayesian approach to the estimation of θ_t naturally allows one to incorporate several sources of uncertainty and produce straightforward estimates of the q.o.i. from the posterior distribution. This section should serve as an up-to-date account of the state-of-the-art. See Branscum et al. (2005) for a review of older Bayesian methods.

As a starting point, consider the model

$$\gamma \sim \text{Beta}(\alpha_\gamma, \beta_\gamma), \tag{7.13}$$
$$\delta \sim \text{Beta}(\alpha_\delta, \beta_\delta),$$
$$\theta_t \sim \text{Beta}(\alpha_\theta, \beta_\theta),$$
$$x \sim \text{Binomial}(n, p),$$

with p given as in (7.9). This model accommodates uncertainty about test characteristics (sensitivity and specificity) whilst allowing for the incorporation of prior information about θ_t. Elicitation of the joint prior on (δ, γ), π_{DG}, is straightforward if one has information of the form in Table 7.1. This information is usually released by test manufacturers.

For simplicity, assume that $\pi_{DG}(\delta, \gamma) = \pi_D(\delta)\pi_G(\gamma)$ (see Section 7.3.7, however). Further, assume uniform —Beta(1, 1)— distributions on δ, γ. Then the distributions in the model in (7.13) can be seen as posterior distributions with $\alpha_\delta = \text{TP} + 1$, $\beta_\delta = \text{FN} + 1$ and $\alpha_\gamma = \text{TN} + 1$, $\beta_\gamma = \text{FP} + 1$. Sometimes researchers are only able to measure (TP, FN) and (TN, FP) in separate essays, that is, not able to measure sensitivity and specificity jointly. This is a minor complication that changes very little in the elicitation procedure. See Section 2 of Gelman and Carpenter (2020) for an analysis with separate sensitivity and specificity data. There might be situations where the actual counts from a validation experiment are not available. If the mean sensitivity (specificity) is reported along $100 \times \alpha\%$ uncertainty intervals, hyperparameters can more often than not be approximated by a simple optimisation procedure that attempts to find the hyperparameters that yield mean and quantiles close to the measured values. If one only has access to the mean $E[\delta] =: m_\delta$, say, our advice is to find the Beta distribution with the highest entropy under the constraint that $\alpha_\delta/(\alpha_\delta + \beta_\delta) = m_\delta$—likewise for γ.

The focus on careful elicitation stems from two main reasons: (i) the need to properly propagate uncertainty about test characteristics and (ii) the fact that the likelihood contains no information about δ and γ. These parameters need to be given strong priors in a Bayesian analysis. Hence, the posterior

$$\xi(\theta_t, \gamma, \delta \mid x, n) \propto p^x (1-p)^{n-x} \pi_{DG}(\delta, \gamma) \pi_P(\theta_t), \qquad (7.14)$$
$$\propto p^x (1-p)^{n-x} \pi_D(\delta) \pi_G(\gamma) \pi_P(\theta_t),$$

with p given by (7.9) is only interesting in the margin $\xi_T(\theta_t \mid x, n)$ since we expect $\xi_D(\delta \mid x, n)$ and $\xi_G(\gamma \mid x, n)$ to closely resemble the π_D and π_G, respectively. The joint posterior (7.14) is intractable and expectations need to be approximated. This is usually accomplished through the use of Markov chain Monte Carlo (MCMC) methods—see Chapter 5 in this book. Whilst the relatively simple structure of the model lends itself to a Metropolis-within-Gibbs scheme that exploits conjugacy where appropriate, recent treatments of this model and its variations (see Section 7.3.6) have employed a general solution using Hamiltonian Monte Carlo (Gelman and Carpenter, 2020), an approach we favour here also.

As far as inferential summaries go, the posterior mean and median are traditional Bayesian point estimates. The construction of credibility intervals deserves a bit more consideration, however. In the beginning of an epidemic caused by a pathogen to which the population has no previous immunity, prevalence will generally be low. As θ_t is constrained to lie in $(0, 1)$, the usual equal-tailed $100 \times \alpha\%$ credibility interval will not be a good inferential summary as it will very likely include 0, which is not reasonable. Gelman and Carpenter (2020) propose using the shortest posterior interval of Liu et al. (2015), which in the case of a unimodal, univariate distribution such as $\xi_T(\theta_t \mid x, n)$ corresponds to the highest posterior density (HPD) interval. The HPD will usually be tighter and exclude the boundaries whilst still allowing a principled treatment of uncertainty about the quantities of interest.

7.3.3 Re-testing positives

In this section we illustrate how the probability calculus previously discussed can be applied in a slightly more complicated scenario. Since the test is imperfect, one strategy is to do confirmatory tests on the samples which test positive in a triage (two-stage) fashion. For convenience, we will drop the individual-level subscript i in the presentation that follows. Suppose now that we have two tests, $T^{(1)}$ and $T^{(2)}$, but only run $T^{(2)}$ on samples (individuals) for which $T^{(1)} = 1$. Let θ_t be the prevalence as before, and let γ_1, δ_1, γ_2, δ_2 be the specificity and sensitivity of tests $T^{(1)}$ and $T^{(2)}$ respectively.

Let Z be the outcome of a re-testing positives-only strategy, i.e., $Z = 1$ if tests 1 and 2 are both positives, and $Z = 0$ otherwise. As before, define $w := P(Z = 1)$. Marginalising over the relevant latent quantities, we have

$$
\begin{aligned}
w &= P(Z = 1, R = 1) + P(Z = 1, R = 0) \\
&= P(Z = 1 \mid R = 1) P(R = 1) + P(Z = 1 \mid R = 0) P(R = 0) \\
&= P(T^{(2)} = 1, T^{(1)} = 1 \mid R = 1) P(R = 1) \\
&+ P(T^{(2)} = 1, T^{(1)} = 1 \mid R = 0) P(R = 0) \\
&= P(T^{(2)} = 1 \mid T^{(1)} = 1, R = 1) P(T^{(1)} = 1 \mid R = 1) P(R = 1) + \\
&+ P(T^{(2)} = 1 \mid T^{(1)} = 1, R = 0) P(T^{(1)} = 1 \mid R = 0) P(R = 0) \\
&= \delta_2 \delta_1 \theta_t + (1 - \gamma_2)(1 - \gamma_1)(1 - \theta_t),
\end{aligned}
$$

where the last line follows from the assumptions that the two tests are conditionally independent given R. The (Bayesian) analysis of this model proceeds in a similar fashion to what has already been discussed, with informative priors on δ_1, γ_1 and δ_2, γ_2. Let $v = \sum_{i=1}^{n} Z_i$. Then

$$
\begin{aligned}
\gamma_1 &\sim \text{Beta}(\alpha_{g1}, \beta_{g1}), \\
\gamma_2 &\sim \text{Beta}(\alpha_{g2}, \beta_{g2}), \\
\delta_1 &\sim \text{Beta}(\alpha_{d1}, \beta_{d1}), \\
\delta_2 &\sim \text{Beta}(\alpha_{d2}, \beta_{d2}), \\
\theta_t &\sim \text{Beta}(\alpha_\theta, \beta_\theta), \\
v &\sim \text{Binomial}(n, w).
\end{aligned}
\tag{7.15}
$$

This setup leads to more accurate estimates of θ_t, especially if $T^{(2)}$ is more accurate than $T^{(1)}$. Moreover, with widespread testing becoming more common, this is a model variation worth exploring.

7.3.4 Estimating underreporting from prevalence surveys

While θ_t is an interesting quantity to estimate in its own right, another major epidemiological goal is to estimate the number of actual cases that have occurred, which is very unlikely to be correctly captured by official figures. One way to obtain such an estimate is to enact the simple correction $Y_t^{\text{corr}} = M\theta_t$. We might also be interested in estimating the fraction p_d of cases which are detected by epidemiological surveillance (i.e., reported). To this end, the model in (7.13) can be extended with the simple approximate model:

$$\gamma \sim \text{Beta}(\alpha_\gamma, \beta_\gamma),$$
$$\delta \sim \text{Beta}(\alpha_\delta, \beta_\delta),$$
$$\theta_t \sim \text{Beta}(\alpha_\theta, \beta_\theta),$$
$$x \sim \text{Binomial}(n, p),$$
$$p_d \sim \text{Beta}(\alpha_d, \beta_d),$$
$$Y_t \sim \text{Binomial}\left(\lfloor M\theta_t \rfloor, p_d\right), \tag{7.16}$$

where $\lfloor a \rfloor$ is the largest integer less than $a \in \mathbb{R}$, also known as the floor function. The underreporting fraction is thus $1 - p_d$. This model is able to incorporate the uncertainty about δ, γ propagated through θ_t. Many fruitful extensions of this model are possible. For instance, if one has a collection of triplets (Y_{tj}, n_j, x_j) from J locations (countries, states, counties, etc.), one can add a regression component to the probability of detection such as, for instance,

$$\text{logit}(p_{dj}) = \beta_0 + \beta_1 X_{j1} + \ldots + \beta_P X_{jP},$$

where the X_j are relevant predictors for each location, such as gross domestic product (GDP), human development index (HDI), number of hospitals *per capita* and other variables thought to be explanatory of a location's ability to detect cases. Epidemiologically speaking, such a model is useful insofar as it allows one to understand the socioeconomic factors influencing local government's capacity for detecting and reporting the disease, which may be useful when deciding where and how to allocate resources. See Chapter 8 for more on regression models.

7.3.5   Illustration: COVID-19 prevalence in Rio de Janeiro

In this section we will illustrate the application of prevalence estimation methods to a real data set from a serological survey conducted in Brazil by the EPI-COVID study (Hallal et al., 2020b,a). The study is a nationwide survey of 133 sentinel cities for which a random sample of up to 250 households was selected. Individuals were then interviewed and tested for antibodies (IgG and IgM) against SARS-CoV-2 using a finger prick sample. Data were collected in three temporal sampling windows, henceforth called phases: phase 1 took place between the 14th and 21st of May 2020, phase 2 between the 4th and 7th of June 2020 and phase 3 between the 14th and the 21st of June. The data are publicly available at http://www.rs.epicovid19brasil.org/banco-de-dados/. According to the manufacturer of the test used in the EPICOVID study (Wondfo), $\delta = 0.848$ with 95% confidence interval (CI) $(0.814, 0.878)$ and $\gamma = 0.99$ with 95% CI $(0.978, 0.998)$. Using this information, we elicit the approximate hyperparameters $\alpha_\delta = 312$, $\beta_\delta = 49$, $\alpha_\gamma = 234$ and $\beta_\gamma = 1$.

The data, along with prevalence estimates obtained using the methods discussed here, are presented in Table 7.3. We show the naïve estimate along

TABLE 7.3: Prevalence estimates (%) for Rio de Janeiro from the EPICOVID data. We present a naïve estimator, the Rogan-Gladen estimator with correction for (fixed) sensitivity and specificity, along with results from fully Bayesian (FB) estimation (see text for details).

	Phase 1	Phase 2	Phase 3
Data (x/n)	5/243	16/250	22/250
Naïve (CI)	2.1 (1.1, 3.0)	6.4 (4.9, 7.9)	8.8 (7.0, 10.6)
Rogan-Gladen (CI)	1.9 (0.8, 3.0)	6.9 (5.1, 8.7)	9.7 (7.7, 11.8)
FB, mean (BCI)	2.3 (0.3, 4.9)	7.3 (4.1, 11.5)	10.1 (6.2, 14.7)
FB, median (HPD)	2.3 (0.0, 4.5)	7.2 (3.9, 11.2)	10.0 (5.9, 14.4)
FB-detection, median (HPD)	2.0 (0.8, 3.5)	6.1 (3.0, 9.3)	8.9 (5.3, 13.1)

with the Rogan-Gladen corrected estimator. Moreover, we show the results for the two fully Bayesian models discussed earlier. For the simpler model in (7.13), we also show the shortest posterior interval (which in this case coincides with the highest posterior density (HPD) interval) discussed by Gelman and Carpenter (2020) along with the more traditional equal-tailed Bayesian credibility interval (BCI). As expected, the HPD is usually tighter than the BCI, although in this example, that difference is unlikely to be of inferential import.

The results for the model with probability of detection in (7.16) are largely in agreement with those from the simpler model, but lead in general to slightly lower estimates for the prevalence. Recall that this model utilises case data and prior information about underreporting and thus in theory contains more information about the process (see below). In agreement with the recent literature (Gelman and Carpenter, 2020; Izbicki et al., 2020), these results make it very clear why it is important to take uncertainty about test characteristics into account.

We have also used the models discussed here to obtain corrected numbers of cumulative infections, I_t^{corr}. In most cases this amounts to a simple rescaling of the estimate for θ_t, that is, $\hat{I}_t^{\text{corr}} = \hat{\theta}_t M$. For the fully Bayesian models, this takes the form of a transformation of the posterior $\xi_T(\theta_t \mid x, n)$. Moreover, the model presented in Section 7.3.4 uses additional information about the probability of detection, p_d. This however necessitates the construction (elicitation) of a prior distribution for p_d. To achieve this, we used data from Wu et al. (2020), who suggest that overall, the number of actual infections is between 3 and 20 times the reported figures in the United States. This suggests that p_d is between 0.05 and 0.334. Covering these two values with 95% probability under a Beta distribution leads to $\alpha_d = 4$, $\beta_d = 20$ and $E[p_d] = 0.16$. These calculations are justifiable in our view because whilst Brazil —and, in particular Rio de Janeiro—and the USA are very different countries, the large variation found by Wu et al. (2020) is likely to encompass the true underreporting in our location of interest.

TABLE 7.4: Estimates of the actual number of cumulative cases in Rio de Janeiro, Y_t^{corr}. We use the same estimators as those in Table 7.3. Notice that for this quantity we report the usual equal-tailed 95% Bayesian credibility interval (BCI). RG = Rogan-Gladen; FB = Full Bayes, FB-det. = Full Bayes with detection probability.

	Phase 1	Phase 2	Phase 3
Cumulative cases (Y_t)	18.7	36.1	53.3
Naïve (CI)	138.8 (77.4, 200.3)	431.9 (327.4, 536.3)	593.8 (472.9, 714.7)
RG (CI)	128.0 (56.6, 199.5)	468.8 (347.3, 590.2)	657.1 (516.5, 797.7)
FB, mean (BCI)	158.5 (22.6, 331.6)	494.7 (275.5, 774.5)	681.2 (415.8, 991.1)
FB-det., mean (BCI)	142.2 (65.7, 256.4)	421.2 (228.2, 658.2)	607.1 (368.5, 901.0)

We estimate the probability (and 95% BCI) of detection in 0.15 $(0.07, 0.28)$ for the first phase, 0.09 $(0.05, 0.16)$ for the second phase and 0.09 $(0.06, 0.14)$ for the third phase. The drop in detection probability could be explained by a surge in cases, which overwhelms the healthcare system and leads to a smaller fraction of cases being reported.

Using notification data up to April 2020 and estimates of the case-fatality ratio (CFR) from the World Health Organisation (WHO), the reporting rate in Rio de Janeiro was estimated by Do Prado et al. (2020) at 0.072 with 95% confidence interval $(0.071, 0.073)$. Our estimates thus encompass those by Do Prado et al. (2020) but incorporate substantially more uncertainty, befitting of a Bayesian analysis.

Table 7.4 shows our estimates of the actual number of cumulative cases, Y_t^{corr}, along with data on the observed cases obtained from `https://brasil.io/home/`. It is important to note how the projections from the fully Bayesian models propagate the uncertainty about model parameters and lead to substantially wider BCIs. Nevertheless, none of the intervals obtained includes the observed number of cases, showing decisively that there is substantial underreporting. Code and data to reproduce the analyses presented here can be found at `https://github.com/maxbiostat/COVID-19_prevalence_Rio`.

7.3.6 Model extensions

We now move on to discuss extensions to the single test model in (7.13) beyond the retesting presented in (7.13). Extensions fall along two main lines: (i) multiple tests and multiple populations and (ii) multilevel (random effects, hierarchical) formulations of the single test model. We shall discuss these in turn.

The first issue one might be confronted with, especially during an epidemic of a new disease, is the absence of a reliable gold standard. If, however, one has access to two (or more) imperfect tests, $T^{(1)}$ and $T^{(2)}$, one can still estimate θ_t by testing n individuals and analysing the resulting contingency table under an appropriate model for the data. One can either assume conditional independence or conditional dependence between $T^{(1)}$ and $T^{(2)}$ and perform Bayesian model selection and averaging to obtain estimates of θ_t that take model uncertainty into account. The choice between conditional dependence or independence is largely problem-specific: if $T^{(1)}$ and $T^{(2)}$ measure different biological processes, then conditional independence is a valid assumption (Branscum et al., 2005). See Black and Craig (2002) for a detailed treatment of the problem of estimating prevalence without a gold standard.

Another issue one might be confronted with is having two tests and two populations with conditional dependence between $T^{(1)}$ and $T^{(2)}$ from which one can use the assumption that (δ_1, γ_1) and (δ_2, γ_2) do not change across populations to alleviate identifiability problems and estimate θ_{t1} and θ_{t2} efficiently. An important observation is that all derivations for model variations of type (i) follow the same basic structure of those presented here: basic conditional probability manipulations. See Branscum et al. (2005) for an excellent treatment of more model variations such as the three tests and k populations model.

We now consider model extensions of type (ii) above: multilevel models. As shown by the illustration in Section 7.3.5, serological surveys are usually performed across locations and time points and their study designs are more often than not non-ignorable. Fortunately, in multilevel (also called random effects or hierarchical) models, the analyst has a powerful tool set at their disposal.

If one has samples from multiple cities from multiple states, one can construct a hierarchical model where there is a general country-wide "effect" which is shared across states, state effects which are shared across cities and city-specific effects for the prevalence. Moreover, one can construct a model which takes spatial dependence into account (see Section 8.5). This may be particularly important in the context of novel a pathogen because disease spread in a naïve population is closely linked to mobility patterns, which have an obvious spatial component. It must be observed that the spatial component in this case cannot be accurately described by a simple nearest-spatial-neighbour structure (Coelho et al., 2020) and that external mobility data must be incorporated into the model in order to construct a meaningful spatial dependence structure.

Another main line of inquiry is taking measurements over time, since prevalence is a dynamic quantity, under the assumption of complete and permanent immunity, θ_t is a monotonically non-decreasing function of time. Dynamic and time series models (cf. Section 8.3) can be leveraged to account for temporal dependence between measurements taken from the target population longitudinally. Flexible priors, such as Gaussian processes, can be employed to include monotonicity and other epidemiologically-motivated constraints (Riihimäki and Vehtari, 2010).

Additionally, spatial and temporal structures can be combined in order to account for complex sampling patterns. A main consideration, however, is computational tractability. The analyst has to balance one hand capturing the sampling structure of the available data with formulating a tractable model for which parameters and other quantities of interest (such as latent variables) are estimable on the other.

A chief application of multilevel models is obtaining projections of θ_t for the whole population from which the sample under analysis was taken. This entails accounting for sample characteristics such as sex, race and socio-economic composition and then projecting what the outcome of interest would be in the general population. In a Bayesian setting, once one has obtained the posterior distribution of model coefficients, one can sample from the posterior predictive distribution using the population characteristics in order to obtain detailed projections of prevalence. This procedure is called multilevel modelling and poststratification (MRP) and has been recently applied to COVID-19 prevalence estimation (Gelman and Carpenter, 2020).

7.3.7 Open problems

Despite (Bayesian) prevalence estimation being a long-standing problem, there are still many open avenues of research. Most approaches, including the one presented in this section, assume independence between a test's sensitivity and its specificity. In reality, this is not a reasonable assumption. Mathematically, we expect δ and γ to be negatively correlated. Thus, explicitly incorporating (prior) dependence between sensitivity and specificity is still an open problem deserving of consideration.

In an epidemic context, temporal dependence is induced by a very specific process: disease transmission. A very important line of investigation is integrating epidemic models of the SIR/SEIR type and prevalence estimation. In addition to explicitly modelling temporal dependence, using epidemic models allows for the estimation of scientifically relevant quantities such as the transmission rate and the effective reproductive number (R_t). More importantly, one can use the fitted models to make epidemiological projections under different scenarios, an almost impossible task with unstructured time-series models. See Larremore et al. (2020) for an application of this framework to COVID-19 in the United States of America.

Finally, we shall address a crucial assumption made throughout this section: that the disease leaves permanent and complete immunity. For COVID-19, for instance, this assumption has been shown to not hold completely. In practice, processes such as antibody waning might interfere with our ability to estimate θ_t precisely if we do not account for the fact that δ and γ are now functions of time. Moreover, there is recent evidence that a certain fraction of asymptomatic individuals never develop detectable levels of antibodies (Milani et al., 2020). It is possible to use the same techniques discussed in this section to derive a model that accounts for a fraction of non-responding individuals. To the best of our knowledge however, this remains an unexplored line of inquiry in the literature.

Acknowledgments

The authors thank the Brazilian National Influenza Surveillance Network (Central Laboratories, National Influenza Centers, state and municipal health secretariats' surveillance teams, and the Influenza Working Group, Department of Immunization and Communicable Diseases of the Health Surveillance Secretariat, Brazilian Ministry of Health) for their partnership.

Bibliography

Barbosa, M. T. S. and Struchiner, C. J. (2002) The estimated magnitude of AIDS in Brazil: A delay correction applied to cases with lost dates. *Cadernos de Saúde Pública*, **18**, 279–285.

Bastos, L. S., Economou, T., Gomes, M. F. C., Villela, D. A. M., Coelho, F. C., Cruz, O. G., Stoner, O., Bailey, T. and Codeço, C. T. (2019) A modelling approach for correcting reporting delays in disease surveillance data. *Statistics in Medicine*, **38**, 4363–4377.

Bastos, L. S., Niquini, R. P., Lana, R. M., Villela, D. A. M., Cruz, O. G., Coelho, F. C., Codeço, C. T., Gomes, M. F. C., Bastos, L. S., Niquini, R. P., Lana, R. M., Villela, D. A. M., Cruz, O. G., Coelho, F. C., Codeço, C. T. and Gomes, M. F. C. (2020) COVID-19 and hospitalizations for SARI in Brazil: A comparison up to the 12th epidemiological week of 2020. *Cadernos de Saúde Pública*, **36**, e00070120.

Black, M. A. and Craig, B. A. (2002) Estimating disease prevalence in the absence of a gold standard. *Statistics in Medicine*, **21**, 2653–2669.

Branscum, A., Gardner, I. and Johnson, W. (2005) Estimation of diagnostic-test sensitivity and specificity through Bayesian modeling. *Preventive Veterinary Medicine*, **68**, 145–163.

Cai, B., Ioannidis, J., Bendavid, E. and Tian, L. (2020) Exact inference for disease prevalence based on a test with unknown specificity and sensitivity. *arXiv preprint arXiv:2011.14423*.

Centers for Disease Control (1988) Guidelines for evaluating surveillance systems. *Morbidity and Mortality Weekly Report*, **37**.

Codeço, C. T., Villela, D. A. M. and Coelho, F. C. (2018) Estimating the effective reproduction number of dengue considering temperature-dependent generation intervals. *Epidemics*, **25**, 101–111.

Coelho, F. C., Lana, R. M., Cruz, O. G., Villela, D. A., Bastos, L. S., Pastore y Piontti, A., Davis, J. T., Vespignani, A., Codeço, C. T. and Gomes, M. F. (2020) Assessing the spread of COVID-19 in Brazil: Mobility, morbidity and social vulnerability. *PloS one*, **15**, e0238214.

Diggle, P. J. (2011) Estimating prevalence using an imperfect test. *Epidemiology Research International*, **2011**.

Do Prado, M. F., de Paula Antunes, B. B., Bastos, L. D. S. L., Peres, I. T., Da Silva, A. D. A. B., Dantas, L. F., Baião, F. A., Maçaira, P., Hamacher, S. and Bozza, F. A. (2020) Analysis of COVID-19 under-reporting in Brazil. *Revista Brasileira de Terapia Intensiva*, **32**, 224.

Flor, M., Weiß, M., Selhorst, T., Müller-Graf, C. and Greiner, M. (2020) Comparison of Bayesian and frequentist methods for prevalence estimation under misclassification. *BMC Public Health*, **20**, 1–10.

Gamerman, D. and Lopes, H. F. (2006) *Markov Chain Monte Carlo Stochastic Simulation for Bayesian Inference*. New York: Chapman and Hall/CRC, 2nd edn.

Gelman, A. and Carpenter, B. (2020) Bayesian analysis of tests with unknown specificity and sensitivity. *Journal of the Royal Statistical Society: Series C (Applied Statistics)*, **69**, 1269–1283. URLhttps://rss.onlinelibrary. wiley.com/doi/abs/10.1111/rssc.12435.

Greenland, S. (1996) Basic methods for sensitivity analysis of biases. *International Journal of Epidemiology*, **25**, 1107–1116.

— (2009) Bayesian perspectives for epidemiologic research: III. Bias analysis via missing-data methods. *International Journal of Epidemiology*, **38**, 1662–1673.

Hallal, P. C., Barros, F. C., Silveira, M. F., Barros, A. J. D. d., Dellagostin, O. A., Pellanda, L. C., Struchiner, C. J., Burattini, M. N., Hartwig, F. P., Menezes, A. M. B. et al. (2020a) EPICOVID19 protocol: Repeated serological surveys on SARS-CoV-2 antibodies in Brazil. *Ciência & Saúde Coletiva*, **25**, 3573–3578.

Hallal, P. C., Hartwig, F. P., Horta, B. L., Silveira, M. F., Struchiner, C. J., Vidaletti, L. P., Neumann, N. A., Pellanda, L. C., Dellagostin, O. A., Burattini, M. N. et al. (2020b) SARS-CoV-2 antibody prevalence in Brazil: Results from two successive nationwide serological household surveys. *The Lancet Global Health*, **8**, e1390–e1398.

Höhle, M. and an der Heiden, M. (2014) Bayesian nowcasting during the STEC O104:H4 outbreak in Germany, 2011. *Biometrics*, **70**, 993–1002.

Izbicki, R., Diniz, M. A. and Bastos, L. S. (2020) Sensitivity and specificity in prevalence studies: The importance of considering uncertainty. *Clinics*, **75**.

La Marca, A., Capuzzo, M., Paglia, T., Roli, L., Trenti, T. and Nelson, S. M. (2020) Testing for SARS-CoV-2 (COVID-19): A systematic review and clinical guide to molecular and serological in-vitro diagnostic assays. *Reproductive Biomedicine Online*, **41**, 483–499.

Lana, R. M., Coelho, F. C., Gomes, M. F. d. C., Cruz, O. G., Bastos, L. S., Villela, D. A. M. and Codeço, C. T. (2020) The novel coronavirus (SARS-CoV-2) emergency and the role of timely and effective national health surveillance. *Cadernos de Saúde Pública*, **36**, e00019620.

Larremore, D. B., Fosdick, B. K., Bubar, K. M., Zhang, S., Kissler, S. M., Metcalf, C. J. E., Buckee, C. and Grad, Y. (2020) Estimating SARS-CoV-2 seroprevalence and epidemiological parameters with uncertainty from serological surveys. *medRxiv*.

Liu, Y., Gelman, A. and Zheng, T. (2015) Simulation-efficient shortest probability intervals. *Statistics and Computing*, **25**, 809–819.

Mack, T. (1993) Distribution-free calculation of the standard error of chain ladder reserve estimates. *ASTIN Bulletin: The Journal of the IAA*, **23**, 213–225.

Martins, T. G., Simpson, D., Lindgren, F. and Rue, H. (2013) Bayesian computing with INLA: New features. *Computational Statistics & Data Analysis*, **67**, 68–83.

Milani, G. P., Dioni, L., Favero, C., Cantone, L., Macchi, C., Delbue, S., Bonzini, M., Montomoli, E. and Bollati, V. (2020) Serological follow-up of SARS-CoV-2 asymptomatic subjects. *Scientific Reports*, **10**, 1–7.

Ministério da Saúde and Secretaria de Vigilância em Saúde (2020) Guia de Vigilância Epidemiológica Emergência de Saúde Pública de Importância Nacional pela Doença pelo Coronavírus 2019: Vigilância de Síndromes Respiratórias Agudas COVID-19.

Ministério da Saúde, Secretaria de Vigilância em Saúde and Coordenação-Geral de Desenvolvimento da Epidemiologia em Serviços (2019) *Guia de Vigilância Em Saúde*. Brasília: Ministério da Saúde, third edn.

Ministério da Saúde, Secretaria de Vigilância em Saúde and Departamento de Vigilância das Doenças Transmissíveis (2018) Plano de Contingência para Resposta às Emergências de Saúde Pública: Influenza – Preparação para a Sazonalidade e Epidemias.

Ministério da Saúde, SINAN (2012) Ficha de Registro individual destinada para unidades com internação. Síndrome Respiratória Aguda Grave (SRAG) - internada ou óbito por SRAG.

Noufaily, A., Farrington, P., Garthwaite, P., Enki, D. G., Andrews, N. and Charlett, A. (2016) Detection of infectious disease outbreaks from laboratory data with reporting delays. *Journal of the American Statistical Association*, **111**, 488–499.

Parikh, R., Mathai, A., Parikh, S., Sekhar, G. C. and Thomas, R. (2008) Understanding and using sensitivity, specificity and predictive values. *Indian Journal of Ophthalmology*, **56**, 45.

Renshaw, A. E. and Verrall, R. J. (1998) A stochastic model underlying the chain-ladder technique. *British Actuarial Journal*, **4**, 903–923.

Riihimäki, J. and Vehtari, A. (2010) Gaussian processes with monotonicity information. In *Proceedings of the Thirteenth International Conference on Artificial Intelligence and Statistics*, 645–652.

Rogan, W. J. and Gladen, B. (1978) Estimating prevalence from the results of a screening test. *American Journal of Epidemiology*, **107**, 71–76.

Rue, H., Martino, S. and Chopin, N. (2009) Approximate Bayesian inference for latent Gaussian models by using integrated nested Laplace approximations. *Journal of the Royal Statistical Society: Series B (Statistical Methodology)*, **71**, 319–392.

World Health Organization (2006) World Health Organization. Communicable disease surveillance and response systems: Guide to monitoring and evaluating.

— (2013) *Global Epidemiological Surveillance Standards for Influenza*.

Wu, S. L., Mertens, A. N., Crider, Y. S., Nguyen, A., Pokpongkiat, N. N., Djajadi, S., Seth, A., Hsiang, M. S., Colford, J. M., Reingold, A. et al. (2020) Substantial underestimation of SARS-CoV-2 infection in the United States. *Nature Communications*, **11**, 1–10.

8

Hierarchical modelling

Dani Gamerman

Universidade Federal de Minas Gerais/Universidade Federal do Rio de Janeiro, Brazil

Marcos O. Prates

Universidade Federal de Minas Gerais, Brazil

CONTENTS

This chapter describes modelling tools that are not directly drawn from the data but provide useful additional components for pandemic data modelling. The presentation starts from the simplest additional component: regression. It evolves to models that accommodate temporal and spatial variation. Finally, it addresses the issue of joint modelling data from different, but related, regions. The presentation makes use of hierarchical specifications where the full model is obtained by an integrated combination of its components.

8.1 Introduction

This chapter presents a number of alternative models that address relevant extensions to the modelling tools of Part II. They are based on the inclusion of additional layers or levels in the model structure. All additional layers of the model structure are motivated by the observed data features and might also

DOI: 10.1201/9781003148883-8

be qualified as data-driven. They acknowledge the presence of heterogeneity in the observed data and accommodate possible descriptions for them, rather than offering model-based explanations. The description can be deterministic, when the source of variation is related to other known variables and might be specified in regression-like terms. This route is described in Section 8.2.

Stochastic specification of the variation is a useful approach to acknowledge in the model the presence of heterogeneity when the source of variation is not entirely related to other known variables. Incorporation of dynamic components to provide flexibility and better temporal adaptation to changes follows this path. These dynamic components require an additional layer for the specification of how the component evolves over time. Markovian dependence is a useful tool here. Some ideas in this direction are provided in Section 8.3.

Similar ideas can be used for joint modelling of many observational units (e.g., countries) and thus enable borrowing of information across units. This also requires an additional layer for the specification of the relation between these units. This is presented in Section 8.4. This section is called Hierarchical Models to follow current terminology, but all sections of this chapter present some form of hierarchy in their respective model layers.

The models are usually not referred to as hierarchical when the hierarchy is induced by some specific structure in the specification. An important special case is provided by the presence of spatial structure in the data. This issue is addressed separately in Section 8.5 due to the growing relevance of spatial models in Epidemiology.

These ideas are well known in the statistical literature and are frequently pursued to accommodate temporal and spatial variability in a variety of models. This chapter will address their integration with the models described in Part II of this book. Examples are provided in each of the following sections to illustrate the range of modelling possibilities.

8.2 Regression

Regression is a simple and yet very powerful tool in Statistics and data analysis in general. It involves including additional variables to help the explanation of the variation exhibited by the data. It started centuries ago from the simplest possible model where the variation the data of interest was described via a linear predictor of explanatory variables for the mean. In mathematical terms

$$E(y) = \mu \text{ and } \mu = \beta_0 + \beta_1 x_1 + \cdots + \beta_p x_p$$

where y is the data of interest such as counts of an epidemic in a given region and $x_1, \cdots, x_p$ are the explanatory variables. The coefficients $\beta_1, \cdots, \beta_p$

inform about the strength of the relation between the explanatory variables (or covariates) and the variable of interest.

This basic idea has experienced considerable growth in recent decades. It now serves many more purposes than merely explaining the variation of the mean of y in a linear relation. It is now used to contemplate any possible relation between any components of the model, not just the data mean, and variables that are deemed to explain their possible variation. As such, the relation between them no longer needs to be linear and the covariates no longer need to be observed quantities.

The models presented in Part II of the book revolved around the logistic model as the primary source of explanation of the progress of disease counts. Some departure from this structure was presented in Chapter 5 where additional features drawn from the observed data where included. None of them considered the use of additional variables to explain them. This section is devoted to describing how to use this auxiliary information in a regression context for the analysis of pandemic data.

Take the generalised logistic curve as the starting point for modelling counts of cases and deaths. This curve is entirely described by the parameter set $\theta = (a, b, c, f)$. Each one of them was associated with a given feature of the data. For example, a is clearly associated with the magnitude of the counts and c controls the rate of growth of the counts, as explained in Chapter 3.

The question one should ask at this point is: is there any other source of information in the pandemic that can help explain why each of the parameters lies in a given region of its possible values?

For example, counts of cases are directly related to the availability of tests to ascertain the presence of the disease in a given individual. If a given region, be it a continent, a country, a state or a city, does not possess the capacity of widespread testing for the disease, the number of cases will inexorably be limited by the number of tests available. Even more obviously, the number of cases cannot be larger than the population size.

These examples suggest that the scale parameter a might be better described by taking these additional variables into account. There are many ways this relation can be mathematically established. An encompassing representation might be

$$g_a(a_t) = \beta_{a,0} + \beta_{a,1}Test_t + \beta_{a,2}Pop_t, \qquad (8.1)$$

where a_t is the time-varying value of a at time t, g_a is a suitable link function relating the covariates and the object of interest a, $Test_t$ is the number of tests performed at time t, Pop_t is the population size at time t, and $\beta_{a,1}$ and $\beta_{a,2}$ are their respective regression coefficients.

Some comments are in order before proceeding:

- The parameter a that has been fixed over time until now became a time-varying parameter a_t as a result of the temporal changes brought by the covariates $Test_t$ and Pop_t.

- The link function g could, in principle, be any function including the identity but it makes sense to be a function that brings a_t to the entire real line as the predictor in the right-hand side of (8.1) lies typically over the entire line. Since a_t are only restricted to be positive, a standard option in this case is the logarithmic link.

- Complexity of the model grows as the number of parameters increases by as many as the number of covariates. Scalability may be a requirement worth considering, especially in the case of pandemics with results revised daily.

- The model remains exactly as in Part II but for the replacement of the expression of the mean in (3.5) by

$$\mu(t) = \frac{a_t \, c \exp\{ct\}}{[1 + b \exp\{ct\}]^2}.$$

Similar ideas could be applied to other parameters of the model. A good example is the parameter c, which controls the infection rate. There are reasons to believe that social distancing procedures and isolation measures may help reduce this rate. This knowledge can be mathematically reflected in a regression model in the form

$$g_c(c_t) = \beta_{c,0} + \beta_{c,1} SD_{1,t} + \cdots + \beta_{c,q} SD_{q,t}, \tag{8.2}$$

where c_t is the time-varying value of c at time t, g_c is a suitable link function relating the covariates and the object of interest c, $SD_{1,t}, \cdots, SD_{q,t}$ are the values of q different social distancing measures at time t and $\beta_{c,1}, \cdots, \beta_{c,q}$ are their respective regression coefficients. The comments above related to the regression for a are equally relevant here in the regression model for c.

Although the parameters a and c are the more easily interpretable model parameters, nothing prevents building regression models for b and f should one find relevant variables to explain their possible variation over time.

The models are completed with a (joint) prior distribution for all the regression coefficients β_js and their intercepts β_0s, replacing the prior distribution for the model parameters for which their regression was built. Typical choices involve normal distributions with 0 mean and large variances, but substantive prior knowledge may force changes in these hyperparameter settings.

Once the regression models are defined, the likelihood function for all model parameters are obtained by replacing (a, b, c, f) with their regression expressions. Combination of the likelihood with the prior density via Bayes' theorem leads to the posterior density. The expression for this latter density is analytically available but for the normalising constant. The expression of this constant will not be available analytically and any of the required summaries of the posterior distribution such as posterior means and (marginal) credibility interval limits will also not be available. Once again, the approximating techniques are used to provide these numerical summaries.

These summaries may help in understanding the relevance of the regression components. The most basic test is the verification of significance of a regression. Evidence of this significance is provided by evaluation of the relative weight (assessed via their posterior distribution) of non-null values for the coefficients, just as in any regression. Tests for evaluation of the relevance of regressions are mentioned in Chapter 6 and fully described in Migon et al. (2014). For example, if the test indicates that the regression coefficients $(\beta_{a,1}, \cdots, \beta_{a,q})$ are irrelevant, then the regression could be discarded and a be assumed to be constant over time.

Finally, there is no reason to restrict regressions to the components of θ. Regressions can be built to explain other relevant features, generically denoted here by ξ. Examples include any of the model features defined in Chapter 3 such as the total number of cases or the pandemic peak time. One possible regression for ξ might be

$$g_\xi(\xi_t) = \beta_{\xi,0} + \beta_{\xi,1} x_{1,t} + \cdots + \beta_{\xi,q} x_{q,t}, \tag{8.3}$$

where the quantities in (8.3) are similar to those used in (8.1) and (8.2).

Other variables that may be brought in to explain a variety of model components are provided by immunisation campaigns. Vaccine coverage over the population and efficiency are some examples of covariates that should have an important effect on model components. Of course their effects will only be appropriately captured by the likelihood after a substantial amount of information is collected over a number of days.

This regression could be brought into the model with other regressions to other model components. One must only avoid having more regressions than the number of identifiable parameters in the model. Failing to do that might bring into the model redundant components that will blur the interpretation even for components that were otherwise well identified.

8.3 Dynamic models

Any pandemic or epidemic is expected to be subject to changes during the course of its occurrence. Regression models described in Section 8.2 show possible forms to accommodate the explanation for these changes. They assume that the analyst knows what causes the changes, which variables quantify the causal agents, the form in which these variables affect the model describing the pandemic and the availability of these variables. There are many situations when there is no clear knowledge of one of the assumptions above. In these situations one is faced with knowledge that the model is varying across a given dimension but does not know the form of this variation.

This section delves into model specifications that address variation over time without perfect knowledge of the source or explanation of this variation.

Variation over other dimensions could also be entertained. Spatial variation across regions lies among the most common dimensions across which changes are expected. This kind of variation is addressed in Section 8.5.1.

The framework under which time-varying modelling tools are developed are usually referred to as dynamic or state-space models. These models rely on the assumption that given model component(s) are best described by a temporal evolution in the form

$$\xi_t = h(\xi_{t-1}, w_t), \tag{8.4}$$

where the model component ξ became time-varying and is indexed by time, h is an appropriately chosen function and w_t is a disturbance term. Equation (8.4) merely establishes a connection between successive values of ξ. This last term, w_t, brings a stochastic element into the evolution to represent the lack of precise knowledge about the temporal variation. Were $w_t = 0$, the evolution would be entirely deterministic and could, in principle, be contemplated in the regression framework of the previous section. It is typically assumed that these terms are independent and identically distributed and are also independent of the previous values of ξ.

An example of a dynamic model is obtained by going back to the examples of the previous section. There, good reasons to assume the parameter a varies over time were presented. But the mathematical representation of this variation might not be known. All that the analyst might be prepared to assume is that successive values of a should be similar as

$$a_t = a_{t-1} + w_t, \text{ where } E(w_t) = 0 \text{ and } Var(w_t) = W_a > 0. \tag{8.5}$$

Note from (8.5) that $E(a_t|a_{t-1}) = E(a_{t-1} + w_t|a_{t-1}) = a_{t-1} + E(w_t|a_{t-1}) = a_{t-1} + E(w_t) = a_{t-1}$, where the independence between the disturbances and previous values of a was used. The above calculation shows that $E(a_t|a_{t-1}) = a_{t-1}$. This local constancy is a reasonable mathematical translation of temporal similarity of the a_t's. It can also be derived that $Var(a_t|a_{t-1}) = Var(a_{t-1} + w_t|a_{t-1}) = Var(a_{t-1}) + Var(w_t|a_{t-1}) = Var(a_{t-1}) + W_a > Var(a_{t-1})$. This enlarged variance reflects the increase in uncertainty as time passes.

The difficulty found in an evolution such as (8.5) is that w_t could be negative and thus may allow $a_t < 0$ with positive probability, even when $a_{t-1} > 0$. This result would lead to an invalid evolution because the values of a at all times should be positive. This possible model imperfection can be easily solved by assuming the evolution for values of a suitably transformed through a link function g. Taking $g = \log$ leads to a transformed evolution

$$\log a_t = \log a_{t-1} + w_t, \text{ where } E(w_t) = 0, \tag{8.6}$$

or equivalently

$$a_t = a_{t-1} w_t^*, \text{ where } w_t^* = \exp\{w_t\}. \tag{8.7}$$

Note that $E(\log a_t | \log a_{t-1}) = \log a_{t-1}$, maintaining the desired local constancy, albeit in the logarithmic scale. If the local constancy should be placed over a rather than over $\log a$, then the required specification for the disturbance must be that $E(w_t^*) = 1$, because $E(a_t | a_{t-1}) = E(a_{t-1} w_t^* | a_{t-1}) = a_{t-1} E(w_t^* | a_{t-1}) = a_{t-1} E(w_t^*) = a_{t-1}$.

Of course, the disturbance terms bring additional uncertainty over the system through their variances $Var(w_t) = W_a$. This added variation quantifies the amount of expected similarity. The larger (smaller) the value of W_a, the less (more) similar the successive values of a_t are expected to be. The variance W_a could be previously set but this is not an easy elicitation task. It might also by indirectly specified through the notion of discount factors as extensively explained in West and Harrison (1997). Nowadays, it is no longer difficult to estimate W_a using the approximation techniques presented in Chapter 6, even though this task brings an additional computational burden.

A probabilistic model for the disturbances must be specified in order to perform Bayesian calculations. The most obvious specifications are based on the normal and Gamma distributions. Then, one possible specification for w_t is a $N(0, W_a)$ distribution. This implies local constancy of the $\log a_t$'s, but would also imply that w_t^* has log-normal distribution with mean $E(w_t^*) = \exp\{W_a/2\} \neq 1$, invalidating the local constancy of the a_t's in their original scale. If the local constancy at the original a scale is desired, then $E(w_t^*)$ must be equal to 1. This is obtained by taking $w_t \sim N(-W_a/2, W_a)$. Specification of values for hyperparameter W_a is not trivial and may be case specific.

The disturbance variances Ws play the important role of informing the model about how fast past information becomes obsolete. This value might be set at a fixed known value as explained above but might also be estimated. Also, it might change over time to indicate relevant pattern variations. Examples of these pattern variations include the onset of a new pandemic wave or the introduction of immunisation campaigns. These data features are likely to introduce changes that could reduce the relevance of past information, when compared to regular evolution. The term regular pandemic is loosely used to describe evolutions remaining in the same mode and without strong external interventions. Examples of intervention include nationwide lockdowns or vaccination campaigns.

These examples would typically force the disturbances to reduce their variances. The amount of decline can be arbitrarily set to a given value, using the notion of discount factors to help. But it can also be estimated, conditional on knowledge of the times that the pattern variation occurs.

Another possible evolution specification for the distribution of the disturbances is to directly model the w_t^* and force them to have mean 1. This is obtained via a $G(W_a, W_a)$ distribution. This distribution also makes it somewhat easier to specify the value of the hyperparameter W_a. Suppose a large probability of a_t and a_{t-1} being close is assumed. For example, one may wish that successive values of a_t differ by no more than 5%, with 90% probability.

Then

$$0.9 = P(0.95 \leq a_t/a_{t-1} \leq 1.05) = P(0.95 \leq w_t^* \leq 1.05),$$

giving $W_a \doteq 1100$. To have the same control over relative difference with .95 probability gives $W_a \doteq 1500$. Similar calculations could be applied to a normal-based evolution but the asymmetry of the log-normal complicates calculations based on symmetric considerations, as in the numerical example above.

Another possibility worth mentioning is provided by the class of dynamic models with exact marginal likelihood described in Gamerman et al. (2013). The class is based on a scaled Beta distribution for the disturbances. When combined with a Gamma prior distribution at the initial time $t = 1$ for a_1 and with a Poisson likelihood, it allows for exact, analytic integration of the time-varying parameters. The integrated likelihood (obtained after removal of the a_t's) depends only on a small number of parameters. This provides an interesting example of the use of the quadrature rules described in Section 6.2.1 in the context of high dimensionality of the parametric space.

Apart from the model in the last paragraph, all ideas above could be applied to introduce time-varying behaviour to the other model parameters b, c and f. The random walk evolution in the log scale seems a reasonable starting point since all of them are positive quantities. Thus, one might have

$$\log \theta_t = \log \theta_{t-1} + w_t, \text{ where } E(w_t) = 0 \text{ and } Var(w_t) = V_w,$$

where $\theta_t = (a_t, b_t, c_t, f_t)$, $\log(z_1, \cdots, z_r) = (\log z_1, \cdots, \log z_r)$, the disturbance term is now an r-dimensional random vector, and V_w is its covariance matrix. Although possible, it seems unlikely to have all model parameters changing over time and to be able to extract data information to estimate all these time-varying parameters.

Note that $E(\log \theta_t | \log \theta_{t-1}) = \log \theta_{t-1}$, maintaining the desired local constancy, albeit in the logarithmic scale. If the local constancy should be placed over θ rather than over $\log \theta$, then the required specification for the disturbance must be that $E(\exp\{w_t\}) = 1$ for the same reasons stated just after (8.7).

The probabilistic specification now involves multivariate distributions for the random disturbance vector. One straightforward extension is to assume independent univariate disturbances, in the form described earlier in this section. The multivariate distribution would then be obtained via the product rule. Although possible, more involved joint distributions are typically not required.

Full account of dynamic models can be obtained from West and Harrison (1997). A much more concise presentation of the same ideas, models, inference and computational tools is provided in Migon et al. (2005).

The idea of allowing model parameters to vary over time without a parametric regression structure driving the variation is not new, even in the context of pandemic modelling. Dynamic models are one of the possible ways, with temporal variation induced via a stochastic non-parametric fashion. Other non-parametric approaches were also proposed recently. Sun et al. (2021) is one of the many possibilities available.

8.4 Hierarchical models

Hierarchical models are an important statistical class of models that can easily be used to include random effects terms to allow for unobserved variability and/or to include dependence between the parameters. Let y_{ij} be a general representation of a response variable for observations $i = 1, \ldots, I$ and $j = 1, \ldots, J$ with conditional mean $E(y_{ij}|X_{ij}, \theta_i) = \mu_i(j)$. Then, a hierarchical model has the following structure

$$
\begin{aligned}
y_{ij} &\sim \pi(\mu_i(j), \phi) \quad i = 1, \ldots, I \\
g(\mu_i(j)) &= h(X_{ij}, \theta_i) \\
\theta_i &\sim \pi_\theta(\cdot|\zeta),
\end{aligned}
\tag{8.8}
$$

where π and π_θ are adequate probability distributions for y_{ij} and θ_i respectively, g is a link function that connects the conditional mean $\mu_i(j)$ to the regression coefficients θ_i of observation i for all j, X_{ij} are the covariates associated with the ij observation, h is an appropriate function of X_{ij} and θ_i, ζ are hyperparameters for the distribution of the regression coefficient θ_i, and ϕ are possible parameters from the likelihood beyond $\mu_i(j)$.

The structure of Equation (8.8) is very generic and basically covers many models in this family. To demonstrate the flexibility of this family and without loss of generality, let us focus at the COVID-19 pandemic example in different regions. In this case, let y_{st} be the daily counts of the number of new cases or deaths at location s and time t. This is equivalent in Equation (8.8) to letting $i = s$ represent the locations and $j = t$ represent the time. Now, combining Equation (8.8) by substituting π by the Poisson distribution, g for the logarithmic link function, and h by the logarithm of Equation (3.8), we have

$$
\begin{aligned}
y_{st} &\sim Poisson(\mu_s(t)) \quad t = 1, \ldots, T, \quad s = 1, \ldots, S \\
\log(\mu_s(t)) &= \log(a_s c_s f_s) - c_s t - (f_s + 1)\log(b_s + \exp(-c_s t)) \\
\theta_s &\sim \pi_\theta(\cdot|\zeta),
\end{aligned}
\tag{8.9}
$$

where $\theta_s = (a_s, b_s, c_s, f_s)^\top$, T is the final observed time and S is the number of locations.

In the next sections, we will show model alternatives to incorporate random effects into modelling to accommodate unknown model variation and how different choices of π_θ can be used with the same purpose.

8.4.1   Unstructured component

As previously discussed, in many situations the analyst does not know or is not capable of specifying the form or the causal agents that determine the observed variability in the data. In this case, let $a_s = a$, $b_s = b$, $c_s = c$, $f_s = f$ be common parameters for all locations. A simple way to incorporate extra variability in the model is the inclusion of independent location-specific random effects. This can be done by changing the second level of Equation (8.9) by

$$
\begin{aligned}
\log(\mu_s(t)) &= \log(acf) - ct - (f+1)\log(b + \exp(-ct)) + \varepsilon_s \\
\theta &\sim \pi_\theta(\cdot|\zeta) \\
\varepsilon_s &\overset{iid}{\sim} N(-\sigma_\varepsilon^2/2, \sigma_\varepsilon^2) \text{ for } s = 1, \ldots, S,
\end{aligned}
\tag{8.10}
$$

where $\theta = (a, b, c, f)$. Notice that a fourth level is included. In this level, an unstructured distribution is assigned to ε_s allowing each area s to have extra variability. The common parameter σ_ε^2 is responsible for measuring the overall extra variability in the regions. In other words, this new model has a random effect (ε_s) that is an unstructured random intercept by location. The inclusion of the random intercept in the logarithm scale of $\mu_s(t)$ leads to local constancy in the transformed scale. However, as pointed out in Equations (8.6) and (8.7), this is equivalent to a multiplicative effect in the original scale. Thus, local constancy in the original scale can be obtained by selecting the prior $N(-\sigma_\varepsilon^2/2, \sigma_\varepsilon^2)$, as discussed in Section 8.3.

Another possibility is to include the extra variability, not as an overall intercept, but to include it specifically in one or more parameters. This can be done by, for example, including a prior in parameter a_s, while letting $b_s = b$, $c_s = c$, $f_s = f$ as

$$
\begin{aligned}
\log(\mu_s(t)) &= \log(a_s cf) - ct - (f+1)\log(b + \exp(-ct)), \\
\theta_{-a} &\sim \pi_{\theta_{-a}}(\cdot|\zeta) \\
\log(a_s) &\overset{iid}{\sim} N(\mu_a, \sigma_a^2) \text{ for } s = 1, \ldots, S,
\end{aligned}
\tag{8.11}
$$

where θ_{-a} is the vector θ without component a, σ_a^2 is the parameter variance, and μ_a is the overall level of the logarithmic transformation of parameter a_s. Notice that μ_a can be either known, unknown, or a regression function

$$
\mu_{a,s} = \beta_{\mu_a,0} + \beta_{\mu_a,1} z_{1,s} + \ldots + \beta_{\mu_a,q} z_{q,s}.
\tag{8.12}
$$

Likewise for ε_s, if μ_a in Equation (8.11) is considered known and set as $-\sigma_a^2/2$, it will guarantee local constancy in the original scale of parameter a_s. On the other hand, an analyst can opt to estimate the overall magnitude by including a new level in the hierarchy and a prior distribution for μ_a. Finally, as indicated in Equation (8.12), another option is to do a regression into relevant explanatory covariates $z_s = (z_{1,s}, \ldots, z_{q,s})$ for $s = 1, \ldots, S$. Clearly, this extension can be done to the other parameters (b, c, f) or to any combination of them.

A perceptive reader will notice that from Equation (8.10) we have that

$$
\begin{aligned}
\log(\mu_s(t)) &= \log(acf) - ct - (f+1)\log(b + \exp(-ct)) + \varepsilon_s \\
&= \log(a) + \log(cf) - ct - (f+1)\log(b + \exp(-ct)) + \varepsilon_s \\
&= \log(cf) - ct - (f+1)\log(b + \exp(-ct)) + (\varepsilon_s + \log(a)) \\
&= \log(cf) - ct - (f+1)\log(b + \exp(-ct)) + \log(a_s).
\end{aligned}
$$

Therefore, under this mean parametrisation, Equations (8.11) and (8.10) are equivalent. Notice that this equivalence is valid only for parameter a and not for the others.

A more complex alternative is to specify a joint prior distribution for two or more parameters. This way, the model is unstructured among the locations but it starts to have structure within each location since the parameters' prior is not independent anymore. Equation (8.13) shows one possible example of how this can be done.

$$
\begin{aligned}
\log(\mu_s(t)) &= \log(ac_s f) - c_s t - (f+1)\log(b_s + \exp(-c_s t)) \\
\theta_{-(b,c)} &\sim \pi_{\theta_{-(b,c)}}(\cdot|\zeta) \\
(\log(b_s), \log(c_s)) &\stackrel{iid}{\sim} N_2((\mu_b, \mu_c), D) \text{ for } s = 1, \ldots, S,
\end{aligned} \tag{8.13}
$$

where D is a 2×2 covariance matrix responsible for capturing the dependence between parameters b_s and c_s, and (μ_b, μ_c) are the mean levels for $(\log(b_s), \log(c_s))$. Similar to the univariate case, the mean levels can be known, unknown or regressed into covariates.

Recently, Schumacher et al. (2021) followed these lines to model the COVID-19 pandemic using series of counts from different countries. Like in Equation (8.13), the authors believed that b_s and c_s are not independent. Therefore, they set a joint distribution of their logarithm as a multivariate skew-t instead of a multivariate normal. As shown by them, the usage of this type of structure allows for the model to improve the parameters estimation. This improvement is a consequence of the enhancement accomplished by learning the parameters' association by using pandemic series in different evolution stages, e.g., the series of different countries. This modelling has been shown to improve fit and to make future predictions more reliable. Another subtle difference is that, instead of assuming a Poisson distribution for the daily death counts of the studied countries, they performed a transformation in the observations y_{st} and assumed that they follow a scale mixture of skew normal distribution (Branco and Dey, 2001).

Table 8.1 shows the parameter estimates, total estimated deaths (TED) and estimated peak dates (EPD) for Brazil, Chile, Italy, Peru, the United States (US), and the United Kingdom (UK) for the daily number of deaths until the 24th of June of 2020. Their code is freely available at the NLME-COVID19 GitHub repository (`https://github.com/fernandalschumacher/`

TABLE 8.1: Fitted parameters for the skew-t non-linear mixed effects model. The total estimated number of deaths (TED) and the estimated date of the peak (EPD) are also presented.

Country	a	b_s	c_s	f	TED	EPD
Brazil		1.517	0.030		96,003	2020-06-08
Chile		1.592	0.018		37,824	2020-08-04
Italy	302,662,797	1.601	0.060	19.323	34,202	2020-04-01
US		1.494	0.047		130,199	2020-04-22
UK		1.586	0.061		40,747	2020-04-15
Peru		1.656	0.028		17,785	2020-06-16

NLMECOVID19) and was used to run the analysis. From this table one can see that there is a strong right skewness in the data ($f \gg 1$). Also, one can see that even for different countries, b_s and c_s are in the same scale and do not vary much. Finally, by the TED and EPD quantities, it is clear today that the COVID-19 pandemic was much more complex and severe than expected by the model at June 24th of 2020.

Figure 8.1 shows a simplified version from the original figure displayed in (Schumacher et al., 2021) for Brazil, Chile, Peru, and the US. The present version does not contain the confidence bands. The confidence interval estimation requires a long bootstrap simulation and was not run. From this figure we can see that the US was estimated to be close to the end of its pandemic. On the other hand, Brazil and Peru were close to their peak, and Chile was in a more initial stage.

8.5　Spatial models

The previous section showed alternatives to incorporate unstructured dependence to capture extra variability in the model. A reasonable assumption when dealing with disease outbreaks is that near areas should present more similar pandemic patterns than distant ones. Therefore, incorporating dependence between locations is an interesting way to improve model fitting and explanatory capacity. Before extending the hierarchical models to capture spatial dependence, we start by briefly introducing areal spatial models. For a detailed overview of spatial modelling, we refer to Cressie (1991) and Banerjee et al. (2014).

In the spatial statistics field, areal spatial dependence can be incorporated in many ways (Besag, 1974; Besag et al., 1991; Leroux et al., 1999; Rodrigues and Assunção, 2012, to mention a few). Without loss of generality, let us focus on the Conditional Autoregressive (CAR, Besag, 1974), which is the

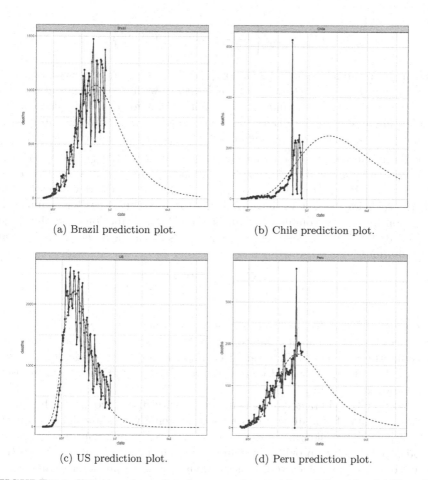

(a) Brazil prediction plot.

(b) Chile prediction plot.

(c) US prediction plot.

(d) Peru prediction plot.

FIGURE 8.1: Fitted and predicted curve, along with real data for (a) Brazil, (b) Chile, (c) US and (d) Peru.

most common areal spatial model used in epidemiology. Let $\delta = (\delta_1, \ldots, \delta_S)$ be a vector of dimension S, the CAR model is specified by the following conditional distributions

$$(\delta_s | \delta_{-s}) \sim N \left(\frac{\rho \sum_{j \sim s} \delta_j}{n_s}, \frac{\sigma_\delta^2}{n_s} \right) \text{ for } s = 1, \ldots, S, \qquad (8.14)$$

where δ_s is the s-th entry of vector δ, δ_{-s} represents the vector δ without component s, $j \sim s$ indicates that regions j and s are neighbours, n_s is the number of neighbours of region s and σ_δ^2 is the overall spatial variability. By defining these conditional distributions, one can show that the joint distribution of δ

is of the form:

$$\pi(\delta) \propto \exp\left\{ -\frac{1}{2\sigma_\delta^2} \delta^\top (D_S - \rho\widetilde{W})^{-1} \delta \right\},$$

where $\widetilde{W}$ is the adjacency matrix composed of zeros and ones, with $\widetilde{w}_{sj} = 1$ if and only if $j \sim s$, D_S is a diagonal matrix with values n_s, and σ_δ^2 is a variance parameter. It is important to notice that for the CAR distribution to be valid one must have $(D_S - \rho\widetilde{W})$ symmetric and positive definite (p.d.). Banerjee et al. (2014) state that if $\rho \in (1/\lambda_{(1)}, 1/\lambda_{(S)})$ where $\lambda_{(1)}$ and $\lambda_{(S)}$ are the minimum and maximum eigenvalues of $D_S^{-1/2}\widetilde{W}D_S^{-1/2}$ respectively, the symmetry and p.d. conditions are satisfied. Moreover, if instead of using $\widetilde{W}$ one uses $W = Diag(1/n_1, \ldots, 1/n_S)\widetilde{W}$ such that each row sums to one in W, then for $|\rho| < 1$ we have that $\Sigma_S^{-1}(\rho) = D_S^{-1}(I - \rho W)$ is symmetric and nonsingular. A common variant of the CAR distribution is the Intrinsic CAR (ICAR, Besag et al., 1991), which yields to the next set of conditional distributions

$$(\delta_s | \delta_{-s}) \sim N\left(\frac{\sum_{j\sim s} \delta_j}{n_s}, \frac{\sigma_\delta^2}{n_s} \right) \text{ for } s = 1, \ldots, S,$$

these conditional distributions are obtained when fixing $\rho = 1$ in Equation (8.14) and are jointly improper. However, as can be seen, it has the nice interpretation that the mean at each location is the average of its neighbours. This distribution will be denoted as $ICAR(W, \sigma_\delta^2)$ and it can be used as a prior in modelling.

Another key characteristic when dealing with disease counts is that areal measurements of this disease can present extreme values and large variability (especially considering rare events in a lower population area). Spatial models are a way to provide smoothing between the areal units, reducing the effect of these possible extreme observations, pulling them closer to an overall center, making model interpretation more realistic and adequate. Of course, suitable smoothing will depend on the specific application and may not be easily quantified.

8.5.1 Spatial component

In this section, we use the hierarchical structure (Section 8.4) to incorporate spatial variability into modelling. First, let $\varepsilon = (\varepsilon_1, \ldots, \varepsilon_S)$ and notice that substituting the third level of Equation (8.10) by

$$\varepsilon \sim N_S\left((-\sigma_\varepsilon^2/2)1_S, \sigma_\varepsilon^2 I_S \right)$$

yields exactly the same model, where 1_S represents the vector of ones with dimension S and I_S is the $S \times S$ identity matrix. Therefore, Equation (8.10) can be rewritten as

$$\begin{aligned}
\log(\mu_s(t)) &= \log(acf) - ct - (f+1)\log(b + \exp(-ct)) + \varepsilon_s \text{ for } s = 1, \ldots, S \\
\theta &\sim \pi_\theta(\cdot|\zeta) \\
\varepsilon &\sim N_S\left((-\sigma_\varepsilon^2/2)1_S, \sigma_\varepsilon^2 I_S\right).
\end{aligned} \tag{8.15}$$

To move from an unstructured random intercept to a spatial one, it is only necessary to change the prior of the vector ε. Thus, if

$$\varepsilon \sim ICAR(W, \sigma_\varepsilon^2)$$

we have that the random intercept is spatially structured. Although this seems to be just a simple change in the prior distribution, it drastically changes the complexity and the interpretation of the models. Now, instead of independent and identically distributed random intercepts, the random intercepts have a multivariate distribution such that the pandemic evolution of all locations are dependent on each other. This model formulation will make $\mu_s(t)$ vary smoothly between neighbouring regions, imposing some similarity in the pandemic pattern in closer areas.

As was done in Section 8.4.1, one can proceed by including the spatial dependence not as a random effect, but incorporating it into the parameters in the model. Again, if this dependence is included in $a = (a_1, \ldots, a_S)$, as shown in Section 8.4.1, the obtained model will be equivalent to the one in Equation (8.15). To exemplify how this can be done, let $a_s = a$, $b_s = b$, $f_s = f$ and $c = (c_1, \ldots, c_s)$, then if

$$\begin{aligned}
\log(\mu_s(t)) &= \log(ac_s f) - c_s t - (f+1)\log(b + \exp(-c_s t)) \text{ for } s = 1, \ldots, S \\
\theta_{-c} &\sim \pi_{\theta_{-c}}(\cdot|\zeta) \\
\log(c) &\sim ICAR(W, \sigma_c^2)
\end{aligned}$$

we have that, in this setting, spatial dependence in the model comes from vector c, which will be spatially dependent and will indirectly impose a spatial structure for the $\mu_s(t)$ for $s = 1, \ldots, S$.

A more complex model can be obtained by including spatial dependence between the areas and also dependence within the parameters. For example, let D be a 2×2 covariance matrix, $\Sigma_S(\rho)$ the covariance matrix of the CAR structure, b and f univariate and a and c vectors of dimension S. A fully dependent model is of the form

$$\begin{aligned}
\log(\mu_s(t)) &= \log(a_s c_s f) - c_s t - (f+1)\log(b + \exp(-c_s t)), \quad s = 1, \ldots, S \\
\theta_{-(a,c)} &\sim \pi_{\theta_{-(a,c)}}(\cdot|\zeta) \\
(\log(a), \log(c)) &\sim N_{2S}(0, \Sigma_S(\rho) \otimes D),
\end{aligned} \tag{8.16}$$

where $\otimes$ represents the Kronecker product. In this simple example, the mean level for the prior was set as 0, but any mean structure like the ones in

Section 8.4.1 for Equation (8.11) can be also adapted. In the spatial literature there are many versions of multivariate spatial models (Gelfand and
Vounatsou, 2003; Carlin and Banerjee, 2003; Jin et al., 2005, 2007; Sain et al.,
2011; MacNab, 2018; Azevedo et al., 2020). Each of these structures offers
different venues that can be used and explored in a more advanced modelling.

Acknowledgements

The chapter authors would like to acknowledge Fernanda Lang Schumacher
for kindly running the NLMECOVID19 analysis and generating Figure 8.1
and Table 8.1 for this chapter.

Bibliography

Azevedo, D. R., Prates, M. O. and Willig, M. R. (2020) Non-separable spatio-
temporal models via transformed Gaussian Markov random fields. *arXiv
preprint arXiv:2005.05464*.

Banerjee, S., Carlin, B. P. and Gelfand, A. E. (2014) *Hierarchical Modeling
and Analysis for Spatial Data*. CRC Press.

Besag, J. (1974) Spatial interaction and the statistical analysis of lattice systems. *Journal of the Royal Statistical Society: Series B (Statistical Methodology)*, **36**, 192–225.

Besag, J., York, J. and Mollie, A. (1991) Bayesian image restoration with two
application in spatial statistics (with discussion). *Annals of the Institute
Statistical Mathematics*, **43**, 1–59.

Branco, M. D. and Dey, D. K. (2001) A general class of multivariate skew-
elliptical distributions. *Journal of Multivariate Analysis*, **79**, 99–113.

Carlin, B. P. and Banerjee, S. (2003) Hierarchical multivariate car models for
spatio-temporally correlated survival data. *Bayesian Statistics*, **7**, 45–63.

Cressie, N. (1991) *Statistics for Spatial Data*. John Wiley & Sons.

Gamerman, D., dos Santos, T. R. and Franco, G. C. (2013) A non-Gaussian
family of state-space models with exact marginal likelihood. *Journal of
Time Series Analysis*, **34**, 625–645.

Gelfand, A. E. and Vounatsou, P. (2003) Proper multivariate conditional autoregressive models for spatial data analysis. *Biostatistics*, **4**, 11–15.

Jin, X., Banerjee, S. and Carlin, B. P. (2007) Order-free co-regionalized areal data models with application to multiple-disease mapping. *Journal of the Royal Statistical Society: Series B (Statistical Methodology)*, **69**, 817–838.

Jin, X., Carlin, B. P. and Banerjee, S. (2005) Generalized hierarchical multivariate car models for areal data. *Biometrics*, **61**, 950–961.

Leroux, B. G., Lei, X. and Breslow, N. (1999) Estimation of disease rates in small areas: A new mixed model for spatial dependence. In *Statistical Models in Epidemiology: The Environment and Clinical Trials* (eds. M. E. Halloran and D. Berry), 179–192. New York: Springer–Verlag.

MacNab, Y. C. (2018) Some recent work on multivariate Gaussian Markov random fields. *Test*, **27**, 497–541.

Migon, H. S., Gamerman, D., Lopes, H. F. and Ferreira, M. A. (2005) Dynamic models. In *Bayesian Thinking* (eds. D. Dey and C. Rao), vol. 25 of *Handbook of Statistics*, 553–588. Elsevier. URLhttp://www.sciencedirect.com/science/article/pii/S0169716105250198.

Migon, H. S., Gamerman, D. and Louzada, F. (2014) *Statistical Inference: An Integrated Approach*. New York: Chapman and Hall/CRC, 2nd edn.

Rodrigues, E. C. and Assunção, R. (2012) Bayesian spatial models with a mixture neighborhood structure. *Journal of Multivariate Analysis*, **109**, 88–102.

Sain, S. R., Furrer, R. and Cressie, N. (2011) A spatial analysis of multivariate output from regional climate models. *The Annals of Applied Statistics*, **5**, 150–175.

Schumacher, F. L., Ferreira, C. S., Prates, M. O., Lachos, A. and Lachos, V. H. (2021) A robust nonlinear mixed-effects model for COVID-19 deaths data. *Statistics and Its Interface*, **14**, 49–57.

Sun, H., Qiu, Y., Yan, H., Huang, Y., Zhu, Y., Gu, J. and Chen, S. (2021) Tracking reproductivity of COVID-19 epidemic in China with varying coefficient SIR model. *Journal of Data Science*, **18**, 455–472.

West, M. and Harrison, J. (1997) *Bayesian Forecasting and Dynamic Models (2nd Ed.)*. Berlin, Heidelberg: Springer-Verlag.

Part IV

Implementation

9

Data extraction/ETL

Marcos O. Prates

Universidade Federal de Minas Gerais, Brazil

Ricardo C. Pedroso

Universidade Federal de Minas Gerais, Brazil

Thaís Paiva

Universidade Federal de Minas Gerais, Brazil

CONTENTS

In this chapter, we present some sources for health data and more specifically COVID-19 repositories. We highlight the importance of official data sources, as well as the necessity of selecting a trustworthy repository to be used. Next, we present the R codes to perform data extraction, transformation and loading (ETL) itself. The transformed data will be used in Chapters 10 and 11 for model fitting and by the online application. This chapter is aimed at R users that would like to perform ETL over data obtained from different sources.

9.1   Data sources

In this chapter, we will describe in detail all the steps of data extraction, transformation and loading (ETL) that are necessary to feed the model and obtain predictions.

Data preparation is a key aspect in statistical analysis. In the era of big data, more and more sources of data are becoming easily available for the general public. First, it is important to identify reliable data sources for the targeted information. Many government agencies

DOI: 10.1201/9781003148883-9

throughout the world freely provide official health data about many aspects of their countries, from social demographics data to actual counts of diseases in different structural levels. For example, the Ministry of Health in Brazil provides detailed public health information at opendatasus.saude.gov.br. Similar information can be obtained in different countries as for the Unites States (cdc.gov), Chile (deis.minsal.cl), Canada (health-infobase.canada.ca), France (santepubliquefrance.fr), German (destatis.de), India (nrhm-mis.nic.in), and many others.

In our case, we are using the COVID-19 data as an example. Therefore, it was important to seek sources that provided not only reliable data but the most up-to-date numbers of cases and deaths. As described in Chapter 2, there are many perspectives to consider regarding notification and occurrence, but for our purpose, focus will be kept on confirmed numbers reported by official government organisations.

For this matter, data can be obtained directly from the different country agencies like in Brazil (covid.saude.gov.br), United Kingdom (coronavirus.data.gov.uk), India (mygov.in/covid-19), France (github.com/opencovid19-fr), Mexico (covid19.sinave.gob.mx) and several others. The ETL of such information, by itself, is a large amount of work and for that reason, it was essential to look for unified data repositories that provide information on a variety of countries and regions. In addition to on-time updates and official original sources, it was also necessary for the repository to be stable and available for automated access. At a global level, some groups created repositories that are daily updated with the most up-to-date information. For example, the Center for Systems Science and Engineering (CSSE) at Johns Hopkins University (github.com/CSSEGISandData/COVID-19), the WHO Coronavirus Disease (COVID-19) Dashboard at the World Health Organisation (covid19.who.int), the European Centre for Disease Prevention and Control (ecdc.europa.eu), the Google Cloud Platform at GitHub (github.com/GoogleCloudPlatform/covid-19-open-data), the Our World in Data organisation (ourworldindata.org) and others. At a national level for Brazil, there are some local repositories that prepare the COVID-19 cases and deaths for different geographical levels as municipalities, states and regions, for example, the *Observatório COVID-19 BR* (github.com/covid19br/covid19br.github.io), the *Brasil IO* (github.com/turicas/brasil.io), the coronabr package (liibre.github.io/coronabr), the covid19br package (Demarqui, 2020) and so on.

Figure 9.1 shows a diagram of the structure of the machinery behind the online application. After selecting repositories that satisfied our conditions: updated daily, automated access, and reliable data source, we prepared an R script that is responsible for downloading the data and performing the necessary transformation to use with the chosen model and load the initial data visualisation.

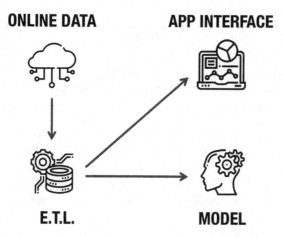

FIGURE 9.1: Diagram representing the data extraction from the repositories, transformation and loading to be accessed by the online application and model fitting.

After looking over the available repositories, we selected the CSSE at Johns Hopkins University for the countries' data, and the *Observatório COVID-19 BR* for the Brazilian data. These repositories have the functionalities we thought were desirable for our application.

9.2 Data preparation

In this section, after selecting the repositories, we present the R codes to prepare the data that will be used by the online application and modelling functions. From the CSSE repository, we obtain the data from the different countries by the following commands in Listing 9.1.

```
1 # set repository address
2 baseURL <- "https://raw.githubusercontent.com/CSSEGISandData/
      COVID-19/master/csse_covid_19_data/csse_covid_19_time_series"
3
4 # download countries files
5 covidworld <- read.csv(file.path(baseURL,"time_series_covid19_
      confirmed_global.csv"), check.names=FALSE,
6     stringsAsFactors=FALSE)
```

Listing 9.1: R code to download the data from the different countries.

Listing 9.2 shows an analogous code to Listing 9.1 to download the Brazilian data from the *Observatório COVID-19 BR* repository.

```
 1  # set repository address
 2  baseURLbr <- "https://raw.githubusercontent.com/covid19br/
       covid19br.github.io/master/dados"
 3
 4  # download states files
 5  covidbr <- read.csv(file.path(baseURLbr,"EstadosCov19.csv"),
       check.names=FALSE, stringsAsFactors=FALSE)
```

Listing 9.2: R code to download the data from the different states in Brazil.

After collecting the raw data from the repositories, it is necessary to prepare and organise it in a data frame that will be used by the online application
to plot the observed data and for the fitted models to perform prediction.

Listings 9.3 and 9.4 show the code to create the data frames that will be
used for a given country and a given Brazilian state, respectively.

```
 1  # For a given country, create the data frame by
 2  # cleaning, renaming and setting initial and final dates
 3
 4  # prepare the data of confirmed cases
 5  covid19_confirm <- covidworld %>%
 6      select(-Lat, -Long) %>%
 7      pivot_longer(-(1:2), names_to="date", values_to="confirmed")
       %>%
 8      mutate(date=as.Date(date, format="%m/%d/%y")) %>%
 9      rename(country='Country/Region', state='Province/State')
10
11  # prepare the data of confirmed deaths
12  covid19_deaths <- read.csv(file.path(baseURL,"time_series_covid19
       _deaths_global.csv"), check.names=FALSE, stringsAsFactors=
       FALSE) %>%
13      select(-Lat, -Long) %>%
14      pivot_longer(-(1:2), names_to="date", values_to="deaths") %>%
15      mutate(date=as.Date(date, format="%m/%d/%y")) %>%
16      rename(country='Country/Region', state='Province/State')
17
18  # join the confirmed and death data into one data frame
19  covid19 <- left_join(covid19_confirm, covid19_deaths,
20      by = c("state", "country", "date"))
21
22  # create the object for one country
23  Y <- covid19 %>% filter(country==country_name) %>%
24    mutate(confirmed_new = confirmed - lag(confirmed, default=0),
25          deaths_new = deaths - lag(deaths, default=0)) %>%
26    arrange(date,state) %>% group_by(date) %>%
27    summarize(n = sum(confirmed, na.rm=T),
28          d = sum(deaths, na.rm=T),
29          n_new = sum(confirmed_new, na.rm=T),
30          d_new = sum(deaths_new, na.rm=T)) %>%
31    arrange(date) %>% filter(date>='2020-01-23')
```

Listing 9.3: R code to download the data from a country of interest.

```
 1  # For a given state, create the data frame by
 2  # cleaning, renaming and setting initial and final dates
 3  Y <- covidbr %>%
```

```
4    rename(name=estado, date=data,
5          cases=casos.acumulados, deaths=obitos.acumulados,
6          new_cases=novos.casos, new_deaths=obitos.novos) %>%
7      mutate(date = as.Date(date)) %>%
8      arrange(name, date) %>% filter(date >= '2020-01-23' &
   date <= last_date & name == state_name) %>%
9      select(date, cases, deaths, new_cases, new_deaths)
```

Listing 9.4: R code to download the data from a state in Brazil.

As will be seen in more detail in Chapter 15, the codes presented here are all encapsulated in the `load_covid()` function from the `PandemicLP` package for more convenient use. To fit the models in our methodology, it is also necessary to have the countries' population, as well as the population for the Brazilian states. The population sizes were obtained from The World Bank (`data.worldbank.org/indicator/SP.POP.TOTL`) and from the *Instituto Brasileiro de Geografia e Estatística* (`www.ibge.gov.br/cidades-e-estados`) for the different countries and Brazilian states, respectively. The population data is available in the online application repository and can be obtained by the code in Listing 9.5. Notice that the variable pop is set to the Brazilian state or country of interest using either line 9 or 12 from Listing 9.5.

```
1  #set repository path
2  popURL = "https://raw.githubusercontent.com/CovidLP/app_COVID19/
      master/R/pop"
3
4  #load the world and Brazilian population files
5  country_pop <- read.csv(file.path(popURL,"pop_WR.csv"))
6  br_pop <- read.csv(file.path(popURL,"pop_BR.csv"))
7
8  #set the pop variable for a state in Brazil
9  pop <- as.numeric(br_pop$pop[which(br_pop$uf == state_name)])
10
11 #set the pop variable for a country
12 pop <- country_pop$pop[which(country_pop$country==country_name)]
```

Listing 9.5: R code to load the population for the Brazilian states or the different countries.

From a practical point of view, it is important to pay attention to some details in the raw data. During the pandemic's course, some countries and states retroactively corrected the total counts of confirmed cases and deaths. This was done by registering a negative number of cases (or deaths) in a given day. For example, Figure 9.2 shows the daily counts of confirmed cases in Spain. It is possible to see the correction in April 24th and May 25th, with negative counts of $10,034$ and 372, respectively.

The models presented in Chapters 3 and 4 require that the number of counts is greater than or equal to zero. To overcome this problem, Listing 9.6 presents a correction over the data to set the negative counts to 0, and remove recursively the counts from the previous days to guarantee the correct total count and that all days satisfy the model assumptions.

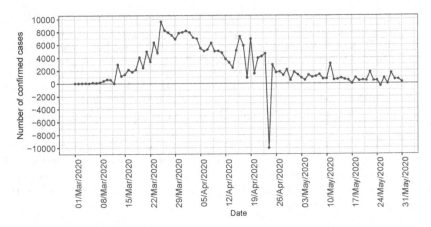

FIGURE 9.2: Daily number of confirmed cases of COVID-19 for Spain in the months of March, April and May.

```r
# performed the count fix for confirmed cases
while(any(Y$new_cases < 0)){
    pos <- which(Y$new_cases < 0)
    for(j in pos){
        Y$new_cases[j-1] = Y$new_cases[j] + Y$new_cases[j-1]
        Y$new_cases[j] = 0
        Y$cases[j-1] = Y$cases[j]
    }
}

# performed the count fix for confirmed deaths
while(any(Y$new_deaths < 0)){
    pos <- which(Y$new_deaths < 0)
    for(j in pos){
        Y$new_deaths[j-1] = Y$new_deaths[j] + Y$new_deaths[j-1]
        Y$new_deaths[j] = 0
        Y$deaths[j-1] = Y$deaths[j]
    }
}
```

Listing 9.6: R code to perform the negative count fix in the confirmed counts or deaths.

From the data visualisation point of view, in the first tab of the online application, we display only the observed time series of confirmed cases and deaths for the selected region, without transforming the negative counts. For this purpose, after the initial processing described in Listings 9.3 and 9.4, the data is ready to be filtered depending on the region selected. For example, if the user selects a specific state, the global data set is filtered to the region selected and only these rows will be displayed in the graph. Otherwise, if the user wants to show the data of the entire country, he can select the input option of all the states, and the data frame will be aggregated to display the country-level values. These data processing steps behind the plots of the app's first tab will be done in a reactive function that will be described in detail in Chapter 11.

After finalising the data extraction and all the preparation described above, the data frame is now ready to be visualised and the predictive analysis performed. In Chapter 10, we present details on how the system is automated to run the models.

9.3 Additional reading

As the reader has noticed, to follow the Implementation part, it is necessary to have some intermediate-level knowledge of R programming. Although we do not introduce R basics in this book, this section aims to guide the reader with initial or no R skills to improve his knowledge, if the objective is to deeply understand the details behind the implementation of an online system.

For beginner users, the Quick-R website (`statmethods.net`) provides step-by-step R scripts to perform a variety of tasks, from R basics and data manipulation, to advanced statistical analysis and visualisation. Besides online material, books are always a good alternative to learn and improve your R skills. In this section, we name a few books that will provide the necessary background to follow and understand our book's Implementation part:

- R in Action (Kabacoff, 2011);

- R Cookbook: Proven Recipes for Data Analysis, Statistics, and Graphics (Teetor, 2011);

- Hands-on Programming with R: Write Your Own Functions and Simulations (Grolemund, 2014);

- R for Data Science: Import, Tidy, Transform, Visualize, and Model Data (Wickham and Grolemund, 2016).

Of course, this list is not exhaustive, and many other books are capable of helping the reader to improve his R proficiency to the level required in ours.

Bibliography

Demarqui, F. N. (2020) *covid19br: Brazilian COVID-19 Pandemic Data*. URLhttps://CRAN.R-project.org/package=covid19br. R package version 0.0.1.

Grolemund, G. (2014) *Hands-on Programming with R: Write Your Own Functions and Simulations*. O'Reilly Media, Inc.

Kabacoff, R. (2011) *R in Action*. Manning Publications, Shelter Island, NY, USA.

Teetor, P. (2011) *R Cookbook: Proven Recipes for Data Analysis, Statistics, and Graphics*. O'Reilly Media, Inc.

Wickham, H. and Grolemund, G. (2016) *R for Data Science: Import, Tidy, Transform, Visualize, and Model Data*. O'Reilly Media, Inc.

10

Automating modelling and inference

Marcos O. Prates

Universidade Federal de Minas Gerais, Brazil

Thais P. Menezes

University College Dublin, Ireland

Ricardo C. Pedroso

Universidade Federal de Minas Gerais, Brazil

Thaís Paiva

Universidade Federal de Minas Gerais, Brazil

CONTENTS

In this chapter, we focus on describing the process of automating the model fitting, and use the COVID-19 example to illustrate how this was done by the CovidLP team behind the results displayed in the online application. We begin by discussing the available software options for performing Bayesian inference. After that, we justify our choice of software and move forward to present the different models that will be fitted and later presented in the online application. Given the different stages of the COVID-19 pandemic around the world, specific models are necessary to adequately fit the observed data. The appropriate model for each country is suggested by the monitoring team among the ones described in Chapters 3, 4 and 5. For the interested reader, we also present the code to parallelise the model fitting and the scheduler script used to automate the process.

DOI: 10.1201/9781003148883-10

10.1 Introduction

With the improvement in computing at the end of the 1980s, Bayesian methods gained more attention and implementing inference using Markov chain Monte Carlo (MCMC) methods became practical for many complex statistical models.

The BUGS (Bayesian Inference Using Gibbs Sampler) project was created in 1989 to provide a flexible tool to fit complex models by using MCMC methods (for a deeper discussion about MCMC methods, see Section 6.2.2). In 1996, the WinBUGS project (Lunn et al., 2000) was created and provided a friendly user interface to perform Bayesian inference in Windows. Next, the project development efforts moved on to the OpenBUGS software (Lunn et al., 2009), which is open source and runs on both Unix/Linux and Windows systems. The JAGS (Just Another Gibbs Sampler) software (Plummer et al., 2003) is an alternative to OpenBUGS. This software is as flexible and also relies on the BUGS language, but it is easily extensible by external contributors, which is not allowed in the BUGS project. For more details about JAGS, see Depaoli et al. (2016). Recently, NIMBLE (Numerical Inference for statistical Models for Bayesian and Likelihood Estimation) (de Valpine et al., 2020) appeared as a faster alternative software to fit flexible models through MCMC. In any case, all the previous software relies on MCMC techniques to perform Bayesian inference which usually requires long chains to achieve convergence in complex models, making them computationally not attractive.

In addition to the aforementioned software, two other options, R-INLA (Rue et al., 2009) and Stan (Gelman et al., 2013), are interesting alternatives for making Bayesian inference. The R-INLA package is built upon the integrated nested Laplace approximation (INLA) methodology. Succinctly, the INLA method approximates the marginal posteriors using Laplace approximations combined with numerical integration. This method avoids sampling and provides reliable posterior approximation with the main advantage of being computationally efficient (usually the fastest option when its use is adequate). However, R-INLA has two main drawbacks: 1) it is built for the class of Latent Gaussian models, which is a wide class but not as flexible as the previous alternatives (e.g., it does not support many types of non-linear models, like the ones used by the CovidLP team); and 2) the user must rely on what is available in the package, because it is not trivial to implement any extension. Finally, instead of relying on traditional MCMC methods, Stan uses the Hamiltonian Monte Carlo (HMC) approach to generate samples. Concisely, the HMC method uses the potential energy of the Hamiltonian system to sample from the posterior distribution (more details about HMC methodology can be seen in Section 6.2.2). It was shown that this approach often converges faster to the posterior distribution than traditional MCMC methods (Hoffman and Gelman, 2014), making it computationally more attractive.

The drawback of **Stan** software is that it requires that all parameters are continuous, and thus does accept discrete parameters. However, for the pandemic models we considered, this is not a problem since all parameters are indeed continuous. Therefore, balancing flexibility and computational performance, we opted to use the **Stan** software for our project. The software has an **R** interface with the **rstan** package (Stan Development Team, 2020), which will be used for Bayesian inference by our group.

When using MCMC methods, as presented in Chapter 6, it is necessary to define the warm-up (to achieve convergence), lag (to reduce correlation within the MCMC sample), the size of the posterior sample and the number of chains. These choices depend on the model complexity and other factors, therefore, it should be tested and calibrated specifically for the application at hand. For the COVID-19 example, after testing, the following choices were made for the parameters:

- **warmup** = 5000 for the generalised logistic and seasonal models, and **warmup** = 8000 for the 2 waves model;

- **lag** = 3;

- **sample_size** = 1000;

- **chains** = 1.

As mentioned in the previous paragraph, usually the HMC used in **Stan** converges faster to the posterior distribution than traditional MCMC methods. Looking at the posterior plots, the potential scale reduction and the effective sample size (see Chapters 6 and 5 for the description of these diagnostic metrics towards convergence), the parametrisation above was selected. For simplicity, the **PandemicLP** package (Chapter 15 and 16) automatically specifies this configuration for the model fitting when the command **covidLPconfig** = **TRUE** in the **pandemic_model** function is applied.

Furthermore, the implementation team in the CovidLP project currently has two machines with 32 cores (64 threads) and 32Gb of RAM each with a Fedora 32 Linux installed as the operational system. Hence, the project can handle up to 128 threads simultaneously, which is a relevant fact to consider when making decisions about which and how many models can be fitted daily for the COVID-19 pandemic and presented in the online application.

Before moving forward with the detailed R scripts, Algorithm 1 presents a conceptual representation of the required steps to automate model fitting. Thus, this pseudo-algorithm aims to help readers to follow the implementation of the R codes presented in the next sections.

Algorithm 1 Pseudo-code of the necessary steps to fit the pandemic model
to time series data.

Set the list of objects (e.g., countries/states);
loop (parallel loop)
    load the data;
    fit the model of interest;
    generate the model prediction;
    generate prediction summary;
    create output list;
    save output to a file;
end loop

10.2 Implementing country models

After having the data prepared by the ETL procedure described in Chapter
9 and selecting **Stan** as our software, we are able to move forward to fit
the different models presented in the online application. The choice of the
appropriate model depends on the pandemic/epidemic current state. We will
explain how to automate and parallelise the model fitting using different stages
of the COVID-19 cases time series across the world (the analysis for the death
counts is equivalent).

The choice of the number of countries to analyse was directly related to
the available computational power of the implementation team (Section 10.1).
This is important to emphasise because one of the aspects decided by the team
was that all of our analysis should be updated and presented daily. Since the
countries data is updated between 04:45 and 05:15 GMT time by the CSSE
repository, and to present the results in the same day, it is important to
emphasise that a CovidLP team decision imposed a restriction that all the
fitted models (or almost all of them) should be updated in the online interface
at 3:00 GMT. Of course, the faster the new results can be released the better.

With this restriction in mind, we started by modelling countries that presented only one wave or were at the beginning of the pandemic considering the
generalised logistic model (Equation (3.8)). As described in Chapter 3, this
model presents important and desirable features to model data that are in
an early stage or those countries that were able to control the pandemic with
a single peak of cases. The package `foreach` (Microsoft and Weston, 2020)
is an alternative to simply implement parallel functions in R. The function
`foreach` combined with the `%dopar%` operator perform a loop that automatically distributes the application of the commands within its brackets among
the registered cores (by the `doMC` package (Analytics and Weston, 2020)). It
allows model fitting for all countries simultaneously.

Listing 10.1 presents the code to fit the one-wave generalised logistic model using the doMC, the foreach and the PandemicLP packages (details of the PandemicLP package functionalities will be presented in Chapters 15 and 16).

```
1  # load packages
2  library(PandemicLP)
3  library(foreach)
4  library(doMC)
5
6  # list of countries provided by the monitoring team
7  countrylist <- c("Argentina","Bolivia","Chile","China",
8                   "Colombia","Ecuador","France","Greece","India",
9                   "Ireland","Korea, South","Mexico","Netherlands",
10                  "New Zealand","Norway","Peru","Paraguay",
11                  "Russia","South Africa","Uruguay","Sweden",
12                  "Switzerland","Turkey","Venezuela","Costa Rica",
13                  "Morocco","Panama","Philippines","Ukraine",
14                  "Ethiopia","Guatemala","Honduras","Indonesia",
15                  "Iraq")
16
17 # register cores
18 registerDoMC(cores = min(63,length(countrylist)))  # number of
      countries or maximum number of cores
19
20 # parallel loop over the countries to fit the model
21 obj <- foreach(s = 1:length(countrylist)) %dopar% {
22
23   # select country
24   country_name <- countrylist[s]
25
26   # load data
27   covid_country <- load_covid(country_name = country_name)
28
29   # fit the model
30   mod <- pandemic_model(covid_country, case_type = "confirmed",
31                         n_waves = 1, covidLPconfig = TRUE)
32
33   # perform model prediction
34   pred <- posterior_predict(mod, horizonLong = 500,
35                             horizonShort = 14)
36
37   # calculate summary statistics
38   stats <- pandemic_stats(pred)
39   stats[[1]] <- NULL # removing the data from 1st position
40   # (the app uses the data coming from other object)
41
42   # rename the lists and data.frame accordingly for the app
43   names(stats) <- c("df_predict", "lt_predict",
44                     "lt_summary", "mu_plot")
45   names(stats$df_predict) <- c("date", "q25", "med", "q975", "m")
46   names(stats$lt_predict) <- c("date", "q25", "med", "q975", "m")
47   names(stats$lt_summary) <- c("NTC25", "NTC500", "NTC975",
48                     "high.dat.low", "high.dat.med",
49                     "high.dat.upper", "end.dat.low",
50                     "end.dat.med","end.dat.upper")
51
52   # prepare the list to be saved
```

```
53    list_out <- list(df_predict = stats$df_predict,
54                     lt_predict=stats$lt_predict,
55                     lt_summary=stats$lt_summary,
56                     mu_plot = stats$mu_plot)
57    name.to.save <- gsub(" ", "-", country_name)
58
59    # saveRDS
60    results_directory <- "/home/CovidLP/Covid/app_COVID19/
         STpredictions/"
61    names(covid_country$data) <- c("date","n","d","n_new","d_new")
62    name.file <- paste0(results_directory,name.to.save,'_',
63                        colnames(covid_country$data)[2],'.rds')
64    saveRDS(list_out, file=name.file)
65
66 }
```

Listing 10.1: R code to fit the generalised logistic model using the `PandemicLP` package. The code is parallelised to fit all the countries in the list simultaneously.

Listing 10.1 can be summarised in the following steps: 1) country list definition (line 7); 2) parallel loop (line 21); 3) load data (line 27); 4) generalised logistic fitting (line 30); 5) model prediction (line 34); and saving the generated output (lines 60–64). This schematic is basically the same for all the models, varying only step 4, where different choices of models can be made depending on the pandemic stage.

We move forward presenting Listing 10.2, which shows how to perform prediction after model fitting. A more interested reader can continue to read in detail the rest of the section, but for the regular user that will rely on the package to fit the model, the discussion and codes presented from this point on may not be of interest and the understanding of Listing 10.1 should be enough. A suggestion for this reader is to check Listings 10.4 and 10.5 and their related paragraphs to learn how to create the bash code to run the script, save the results in the GitHub repository, and how to implement a scheduler to automate the whole process. For advanced readers, we move on by presenting a summary code for the `posterior_predict` function (the full function has more features, details and robustness checking. Details of this function can be obtained in the repository `github.com/CovidLP/PandemicLP` of the `PandemicLP` package).

The first function (lines 1–32) in Listing 10.2 prepares the object fitted by the function `pandemic_model` to create the future predictions. The fitted model has the necessary information, and the auxiliary function `generatePredictedPoints_pandemic` (lines 34–47) is called to perform prediction. This function computes, for each future time, the mean of the process using the generalised logistic function, and then generates samples using the Poisson distribution. Once the samples are created, the output is prepared and returned. The function includes some hidden methods to guarantee a long enough horizon to correctly compute the summary statistics. More data preparation is done and an `output` object is returned.

```
 1  posterior_predict.pandemicEstimated =
 2    function(object, horizonLong = 500, horizonShort = 14,...){
 3
 4    chains = as.data.frame(object$fit)
 5    pop = object$Y$population
 6
 7    M = nrow(chains) # total iterations
 8
 9    finalTime = sum(grepl( "mu", names(chains)) ) # how many mu's
10
11    # generate points from the marginal predictive distribution
12    pred = generatePredictedPoints_pandemic(M,chains,
13                                   max(1000,horizonLong),
14                                   finalTime)
15
16    # create the hidden objects with the 1000 steps ahead
17    methods::slot(object$fit,"sim")$fullPred$thousandLongPred =
         pred$yL # long term prediction
18    methods::slot(object$fit,"sim")$fullPred$thousandShortPred =
         pred$yS + object$Y$data$cases[nrow(object$Y$data)] # observed
         cases + short term prediction
19    methods::slot(object$fit,"sim")$fullPred$thousandMus = pred$mu
         # estimated mean past and future
20
21    y.futL = as.matrix(methods::slot(object$fit,"sim")$fullPred$
         thousandLongPred[,1:horizonLong])
22    y.futS = as.matrix(methods::slot(object$fit,"sim")$fullPred$
         thousandShortPred[,1:horizonShort])
23
24    output <- list(predictive_Long=y.futL, predictive_Short=y.futS,
25                   data=object$Y$data, location=object$Y$name,
26                   cases_type=object$cases.type,
27                   pastMu=as.data.frame(object$fit)[grep("mu",names
         (chains))],
28                   futMu=as.matrix(pred$mu[,1:horizonLong]),
29                   fit=object$fit)
30
31    return(output)
32  }
33
34  generatePredictedPoints_pandemic = function(M,c,h,ft){
35    y = mu = matrix(-Inf,ncol = h,nrow = M)
36
37    # for each future time estimate the mean and generate a sample
38    for (i in 1:h){
39      mu[,i] = exp( log(c$f) + log(c$a) + log(c$c) -
40                    (c$c*(ft+i))-(c$f+1)*log(c$b+exp(-c$c*(ft+i))) )
41      y[,i] = stats::rpois(M,mu[,i])
42    }
43    list(yL = y, yS = t(apply(y,1,cumsum)), mu = mu)
44  }
```

Listing 10.2: Summary of how to perform prediction after model fitting.

After creating samples of the future observations from the model, it is necessary to create the summary statistics that will be provided on the online application. Although the generalised logistic model has closed forms for the peak date, end date, and total number of expected counts (Section 3.3), this is the simplest model used in our analysis and the same analytical results cannot be obtained for all the other models. Therefore, methods to provide empirical versions of these quantities were developed and implemented. These methods have the following advantages: 1) work for the different models used to fit the data, and 2) being able to provide not only point estimates but also credible intervals that are important to quantify the uncertainty in our predictions.

The summarised version of the `pandemic_stats` function is presented in Listing 10.3 (see the `PandemicLP` package repository `github.com/CovidLP/PandemicLP` for more details). The code chunk from lines 10–15 creates the short-term summary. This list returns the date and count quantiles (2.5%, 50% and 97.5%) for the short-term prediction. Next, the total number of expected cases is calculated by summing the observed cases with the samples that provide the median and the 2.5% and 97.5% quantiles (lines 19–27). The next step is to calculate the peak date of the pandemic. We considered as peak the time with the maximum number of daily cases (this definition is necessary for the models with two waves, where there are two local maximums). Therefore, a simple search is done to find the date where the maximum of daily cases is obtained (line 57). The concept of the credible interval requires more care, since it should cover all possible curves that peak within the peak date of the 2.5% quantile curve and the 97.5% quantile curve. To satisfy this condition, we select the two dates such that the 97.5% percentile curve coincides with the highest value of the 2.5% percentile curve (lines 42–56). Finally, code chunk (lines 60–74) calculates the pandemic estimated end date. The pandemic end date is considered to be the date that the 99% percentile of the total counts is achieved. This search is done for the three quantile curves of interest (2.5%, 50% and 97.5%). Lastly, the long-term prediction and summary values are saved in objects to be returned by the function.

```
1  pandemic_stats <- function(object){
2
3    # collect the observed data and the short and long term
       predictions
4    t = length(object$data[[1]])
5    ST_horizon = ncol(object$predictive_Short)
6    LT_horizon = ncol(object$predictive_Long)
7    date_full <- as.Date(object$data$date[1]:(max(object$data$date)
       + max(1000,LT_horizon)), origin = "1970-01-01")
8
9    # list output ST prediction:
10   ST_predict <- data.frame(
11     date  = date_full[(t+1):(t+ST_horizon)],
12     q2.5  = apply(object$predictive_Short,2,stats::quantile,.025),
13     med   = apply(object$predictive_Short,2,stats::median),
14     q97.5 = apply(object$predictive_Short,2,stats::quantile,.975),
15     mean  = colMeans(object$predictive_Short))
16   row.names(ST_predict) <- NULL
17
18   # total number of cases
19   cumulative_y =   object$data$cases[t]
```

```
20
21   lowquant <- apply(methods::slot(object$fit,"sim")$fullPred$
        thousandLongPred,2,stats::quantile,.025)
22   medquant <- apply(methods::slot(object$fit,"sim")$fullPred$
        thousandLongPred,2,stats::median)
23   highquant <- apply(methods::slot(object$fit,"sim")$fullPred$
        thousandLongPred,2,stats::quantile,.975)
24
25   TNC2.5 = sum(lowquant) + cumulative_y
26   TNC50  = sum(medquant) + cumulative_y
27   TNC97.5 = sum(highquant) + cumulative_y
28
29   # calculate the peak and end dates
30   peak2.5 <- peak50 <- peak97.5 <- NULL
31   end2.5 <- end50 <- end97.5 <- NULL
32
33   # get the mean samples from the object
34   chain_mu <- cbind(object$pastMu,
35     methods::slot(object$fit,"sim")$fullPred$thousandMus)
36
37   # quantile of the fitted mean cases
38   mu50 <- apply(chain_mu, 2, stats::quantile, probs = 0.5)
39   mu25 <- apply(chain_mu, 2, stats::quantile, probs = 0.025)
40   mu975 <- apply(chain_mu, 2, stats::quantile, probs = .975)
41
42   # calculate the upper and lower bound dates for the peak
43   mu25_aux <- mu25
44   posMaxq2.5 <- which.max(mu25_aux)
45   aux <- mu975 - mu25_aux[posMaxq2.5]
46   aux2 <- aux[ posMaxq2.5 : length(aux)]
47   val <- ifelse( length(aux2[aux2 < 0]) > 0,
48                  min(aux2[aux2 > 0]), aux[length(aux)] )
49   date_max <- which(aux == val)
50
51   aux <- mu975 - mu25_aux[posMaxq2.5]
52   aux2 <- aux[1:posMaxq2.5]
53   val <- min(aux2[aux2>0])
54   date_min <- which(aux == val)
55
56   # get peak dates
57   peak2.5  <- date_full[date_min]
58   peak97.5 <- date_full[date_max]
59   peak50 <- date_full[which.max(mu50)]
60
61   # calculate the upper and lower bound dates for the end of the
        pandemic
62   q <- .99
63   med_cumulative <- apply(as.matrix(mu50), 2, cumsum)
64   med_percent<- med_cumulative / med_cumulative[t + max(1000,LT_
        horizon)]
65   med_end <- which(med_percent - q > 0)[1]
66   end50 <- date_full[med_end]
67
68   low_cumulative <- apply(as.matrix(mu25), 2, cumsum)
69   low_percent <- low_cumulative / low_cumulative[t + max(1000,LT_
        horizon)]
70   low_end <- which(low_percent - q > 0)[1]
```

```
71   end2.5 <- date_full[low_end]
72
73   high_cumulative <- apply(as.matrix(mu975), 2, cumsum)
74   high_percent <- high_cumulative / high_cumulative[t + max(1000,
        LT_horizon)]
75   high_end <- which( high_percent - q > 0)[1]
76   end97.5 <- date_full[high_end]
77
78   # LT predictions
79   LT_predict = data.frame(date = date_full[(t+1):(t+LT_horizon)],
80                           q2.5 = lowquant[1:LT_horizon],
81                           med = medquant[1:LT_horizon],
82                           q97.5 = highquant[1:LT_horizon],
83                           mean= colMeans(object$predictive_Long))
84   row.names(LT_predict) <- NULL
85
86   # Long-term summary
87   LT_summary <- list(total_cases_LB = TNC2.5,
88                      total_cases_med = TNC50,
89                      total_cases_UB = TNC97.5,
90                      peak_date_LB = peak2.5,
91                      peak_date_med = peak50,
92                      peak_date_UB = peak97.5,
93                      end_date_LB = end2.5,
94                      end_date_med = end50,
95                      end_date_UB = end97.5)
96
97   muplot <- data.frame(date = date_full[1:(t+LT_horizon)],
98                        mu = mu50[1:(t+LT_horizon)])
99   row.names(muplot) <-NULL
100
101  dataplot <- list(data=object$data, location=object$location,
102                   case_type=object$cases_type)
103
104  output <- list(data=dataplot, ST_predict=ST_predict,
105                 LT_predict=LT_predict, LT_summary=LT_summary,
106                 mu = muplot)
107
108  return(output)
109 }
```

Listing 10.3: Summary of how to empirically calculate the peak date, pandemic duration and total counts.

The CovidLP team created a GitHub repository where the online application accesses the results to provide visualisation of results. To automatically run the model fitting script (Listing 10.1, named world_prediction.R), a bash script is created (shown in Listing 10.4). The bash script first pulls the repository to guarantee that it is up-to-date, runs the R script and once finalised, pulls the repository again. This step is important because the model fitting can take from minutes to hours and other models are being fitted simultaneously, e.g., Brazilian states or death cases. Therefore, as a security measure and to avoid any incompatibility with the pull realised before the model fitting, a second round is done just before submission of the obtained results. Next, the script uploads the new files to the GitHub repository. In lines 4, 8 and 14 where the git commands are presented, the **username** and **passwd** must be substituted for the true ones of the repository.

```bash
#!/bin/bash
cd /home/CovidLP/Covid/app_COVID19/

git pull https://username:passwd@github.com/CovidLP/app_COVID19
    master

Rscript /home/CovidLP/Covid/R/STAN/world_prediction.R

git pull https://username:passwd@github.com/CovidLP/app_COVID19
    master
git add /home/CovidLP/Covid/app_COVID19/STpredictions/*

datesub=$(date +%m_%d_%Y_%T)
git commit -m "update results_$datesub"

git push https://username:passwd@github.com/CovidLP/app_COVID19
    master
```

Listing 10.4: Bash code to fit model and upload results to GitHub repository.

Finally, the **Cron** scheduler is used. The **Cron** scheduler allows a Linux user to specify the hour, dates and recurrence that a given code should be executed. The scheduler is changed by the command **crontab -e** and it is presented in Listing 10.5.

```
00:06 * * * /home/CovidLP/Covid/execute_WR.sh
```

Listing 10.5: Scheduler code to call the bash script to fit the generalised logistic model and upload the results to the GitHub repository.

By September of 2020, some countries were already experiencing a second hit of the COVID-19 pandemic, commonly called the "second wave". To fit the two-wave model and automate the process, the steps are exactly the same with a trivial change. Basically, it is necessary to change line 21 of Listing 10.1 to the one presented in Listing 10.6 where the two-wave model (Chapter 5) is selected.

```
mod <- pandemic_model(covid_country, case_type = "confirmed",
                      n_waves = 2, covidLPconfig = TRUE)
```

Listing 10.6: Command line to fit the two-wave model.

It is important to notice that the summary versions of the codes presented in Listings 10.2 and 10.3 were simplified for the generalised logistic model. However, in the **PandemicLP** package, the functions **posterior_predict** and **pandemic_stats** work correctly for all models available in the package. Although code implementation inside the functions may require some adaptation, the baseline idea for all models is the same to the ones presented in this section.

10.3   Implementing Brazilian models

The CovidLP team consists of a group of Brazilian researchers, and for that reason a more detailed analysis for COVID-19 data is done for this country. The analysis of Brazilian data will be used as an example to introduce the seasonal model. It was noticed by the CovidLP monitoring team that there was a common under-notification in the data on Sundays and Mondays for the majority of states in Brazil. Although this fact could be due to many reasons, as discussed in Chapter 2, identifying the true reason and correcting for this fact is not the focus of our project. As the book title emphasises, the project is data-driven and, therefore, we proposed a model to adapt to the characteristics observed in the data.

In this section, we provide a more direct approach of the implementation for the models when predicting the Brazilian data. For an advanced reader interested in more details about the implementation, we refer back to Listings 10.2 and 10.3 for the explanation about the functions' contents. Clearly, the codes presented in the previous section are not from the same model fitted for Brazil. However, the structure behind the functions is the same and rewriting them would be unnecessary and repetitive.

Next, in Listing 10.7, we present all the steps to fit the states' seasonal model. A more keen reader will notice that the structure of the code is equivalent to the one presented in Listing 10.1. As before, the script can be summarised in the following steps: 1) states' list definition (lines 6–7); 2) parallel loop (line 13); 3) load data (line 19); 4) seasonal model fitting with Sunday and Monday selected as the weekdays on which the under-notification occurs (line 23); 5) model prediction (line 29); 6) get the summaries for the online application (line 33); and 7) save the generated output (lines 37–61). The seasonal model is presented in detail in Section 5.3. In summary, this model fits a generalised logistic curve with multiplicative effects on the selected weekdays that allow for under- or over-notifications on those days.

```
1  library(PandemicLP)
2  library(foreach)
3  library(doMC)
4
5  # get the Brazilian state list
6  states <- state_list()
7  state_list <- states$state_abb
8
9  # register cores
10 registerDoMC(cores = min(63,length(state_list)))   # number of
       states or maximum number of cores
11
12 # parallel loop over the countries to fit the model
13 obj <- foreach(s = 1:length(state_list)) %dopar% {
14
15   # set state name
16   state_name <- state_list[s]
```

```
17
18    # load data
19    covid_state <- load_covid(country_name = "Brazil",
20                              state_name = state_name)
21
22    # fit the model
23    mod <- pandemic_model(covid_state, case_type = "confirmed",
24                          n_waves = 1,
25                          seasonal_effect = c("sunday","monday"),
26                          covidLPconfig = TRUE)
27
28    # perform model prediction
29    pred <- posterior_predict(mod, horizonLong = 500,
30                              horizonShort = 14)
31
32    # calculate summary statistics
33    stats <- pandemic_stats(pred)
34    stats[[1]] <- NULL # removing the data (the app uses the data
          coming from other object)
35
36    # rename the lists and data.frame accordingly for the app
37    names(stats) <- c("df_predict","lt_predict","lt_summary",
38                      "mu_plot")
39    names(stats$df_predict) <- c("date", "q25", "med", "q975", "m")
40    names(stats$lt_predict) <- c("date", "q25", "med", "q975", "m")
41    names(stats$lt_summary) <- c("NTC25", "NTC500", "NTC975",
42                                 "high.dat.low", "high.dat.med",
43                                 "high.dat.upper", "end.dat.low",
44                                 "end.dat.med", "end.dat.upper")
45
46    list_out <- list(df_predict = stats$df_predict,
47                     lt_predict=stats$lt_predict,
48                     lt_summary=stats$lt_summary,
49                     mu_plot = stats$mu_plot)
50
51    # saveRDS - Summary files
52    results_directory = "/home/CovidLP/Covid/app_COVID19/
          STpredictions/"
53    names(covid_state$data) <- c("date","n","d","n_new","d_new")
54    file_id <- paste0(state_list[s],'_',
55                      colnames(covid_state$data)[2],'e')
56    saveRDS(list_out,
57            file=paste0(results_directory,'Brazil_',file_id,'.rds'))
58
59    # saveRDS - The posterior predict chains (necessary to create
          the aggregate model of Brazil
60    results_directory = "/home/CovidLP/Covid/app_COVID19/
          STpredictions/"
61    file_id <- paste0(state_list[s],'_posterior_predict_',
62                      colnames(covid_state$data)[2],'e')
63    saveRDS(pred, file = paste0(results_directory,file_id,'.rds'))
64    }
```

Listing 10.7: R code to fit the seasonal model using the PandemicLP package. The code is parallelised to fit all the states in Brazil simultaneously.

As in Listing 10.4, a bash script is created to fit the Brazilian data (Listing 10.8). The bash script performs the same steps: synchronises the repository; fits the models; does a security synchronisation; pushes the files to the GitHub repository. Again, username and passwd must be substituted by its true values (on lines 4, 9 and 15).

```
1  #!/bin/bash
2  cd /home/CovidLP/Covid/app_COVID19/
3
4  git pull https://username:passwd@github.com/CovidLP/app_COVID19
      master
5
6  Rscript /home/CovidLP/Covid/R/STAN/brazil_prediction.R
7  Rscript /home/CovidLP/Covid/R/STAN/aggregate_brazil.R
8
9  git pull https://username:passwd@github.com/CovidLP/app_COVID19
      master
10 git add /home/CovidLP/Covid/app_COVID19/STpredictions/*
11
12 datesub=$(date +%m_%d_%Y_%T)
13 git commit -m "update results_$datesub"
14
15 git push https://username:passwd@github.com/CovidLP/app_COVID19
      master
```

Listing 10.8: Bash code to fit model and upload to the GitHub repository.

Finally, the Cron scheduler is edited (Listing 10.9). With the command crontab -e, a new line is included to also set off the bash script for the Brazilian model (Listing 10.8). Notice that the Brazilian repository is updated earlier and, hence, its task starts before.

```
1  00:03 * * * /home/CovidLP/Covid/execute_BR.sh
2  00:06 * * * /home/CovidLP/Covid/execute_WR.sh
```

Listing 10.9: Scheduler code to call the bash script to fit the models and upload the results to the GitHub repository.

Because the model is fitted for all the Brazilian states, instead of fitting the model one more time for the aggregated data in Brazil, the CovidLP team understood that a more realistic country-level model would be to sum the seasonal models fitted for each state. We believe that this approach is more appropriate because Brazil is a country with continental dimensions, so the pandemic evolutions and stages drastically differ between states and regions. Thus, a sum of the separate models will respect the local characteristics and provide a more realistic fit for the country data than fitting the aggregated data. A theoretical explanation of why using the sum of sub-regions fitted models is a more reliable estimate of the aggregated region is discussed in Section 5.2.

To be able to perform inference by summing the states' models, it is necessary to save the full prediction information beyond its summary (this is done in lines 60–63 of Listing 10.7). With this extra information stored in a

separate file, Listing 10.10 shows the code to aggregate the Brazilian states'
information and construct the inference for the whole country.

```
1  library(PandemicLP)
2
3  # get the Brazilian state list
4  states <- state_list()
5  uf <- states$state_abb
6
7  # directory where the data is stored
8  dir_rds <- "/home/CovidLP/Covid/app_COVID19/STpredictions"
9
10 # read the posterior file for the first state
11 state_nm <- paste0(dir_rds,'/',uf[1],"_posterior_predict_ne.rds")
12 data_base <- readRDS(state_nm)
13 uf_2 <- uf[-1]
14
15 # remove posterior_predict file of the first state from directory
      (to save server space)
16 file.remove(state_nm)
17
18 # get the mean sample and set it to be dates x mcmc sample
19 mu_t <- t(data_base$pastMu)
20
21 # include the dates in the data frame
22 mu_final <- data.frame(data = data_base$data$date,mu_t)
23 names_mu <- names(data_base$pastMu)
24
25 # get hidden objects (necessary for the pandemic_stats function)
26 hidden_short_total <- methods::slot(data_base$fit,"sim")$fullPred
      $thousandShortPred
27 hidden_long_total <- methods::slot(data_base$fit,"sim")$fullPred$
      thousandLongPred
28 hidden_mu_total <- methods::slot(data_base$fit,"sim")$fullPred$
      thousandMus
29
30 # loop for each state - starting with the second one
31 for (u in uf_2) {
32
33   # rds import for the selected state
34   state_nm <- paste0(dir_rds,'/',u,"_posterior_predict_ne.rds")
35   data_uf <- readRDS(state_nm)
36
37   # sum the variables predictive_Long, predictive_Short and futMu
38   data_base$predictive_Long <- data_base$predictive_Long +
39                                data_uf$predictive_Long
40   data_base$predictive_Short <- data_base$predictive_Short +
41                                 data_uf$predictive_Short
42   data_base$futMu <- data_base$futMu + data_uf$futMu
43
44   # create a large data frame by concatenating samples for
        current state in the mean data frame
45   mu_t <- t(data_uf$pastMu)
46   mu_2 <- data.frame(data = data_uf$data$date,mu_t)
47   names_mu <- c(names_mu,names(data_uf$pastMu))
48   mu_final <- rbind(mu_final,mu_2)
49
```

```
50   # merge datasets by date since they can differ on start
51   data_base$data <- merge(data_base$data,data_uf$data,
52                           by = "date", all = TRUE)
53   data_base$data[is.na(data_base$data)] = 0
54   data_base$data$cases.x = data_base$data$cases.x +
55                           data_base$data$cases.y
56   data_base$data$deaths.x = data_base$data$deaths.x +
57                           data_base$data$deaths.y
58   data_base$data$new_cases.x = data_base$data$new_cases.x +
59                           data_base$data$new_cases.y
60   data_base$data$new_deaths.x = data_base$data$new_deaths.x +
61                           data_base$data$new_deaths.y
62   data_base$data <- data_base$data[,-c(6:9)]
63   names(data_base$data) <- c("date","cases","deaths",
64                           "new_cases","new_deaths")
65
66   # sum hidden objects (necessary for the pandemic_stats function
         )
67   hidden_short_uf <- methods::slot(data_uf$fit,"sim")$fullPred$
         thousandShortPred
68   hidden_short_total <- hidden_short_total + hidden_short_uf
69   hidden_long_uf <- methods::slot(data_uf$fit,"sim")$fullPred$
         thousandLongPred
70   hidden_long_total <- hidden_long_total + hidden_long_uf
71   hidden_mu_uf <- methods::slot(data_uf$fit,"sim")$fullPred$
         thousandMus
72   hidden_mu_total <- hidden_mu_total + hidden_mu_uf
73
74   # remove posterior_predict file from directory (to save server
         space)
75   file.remove(state_nm)
76 }
77
78 # create hidden object (necessary for the pandemic_stats function
         )
79 methods::slot(data_base$fit,"sim")$fullPred$thousandShortPred <-
         hidden_short_total
80 methods::slot(data_base$fit,"sim")$fullPred$thousandLongPred <-
         hidden_long_total
81 methods::slot(data_base$fit,"sim")$fullPred$thousandMus <- hidden
         _mu_total
82
83 # aggregate the mean samples
84 mu_final <- aggregate(. ~ data, data=mu_final, FUN=sum)
85 mu_final <- mu_final[,-1]
86 mu_final <- t(mu_final)
87 names_mu <- unique(names_mu)
88 colnames(mu_final) <- names_mu
89 data_base$pastMu <- mu_final
90
91 # calculate summary statistics
92 stats <- pandemic_stats(data_base) # calculate stats
93 stats[[1]] <- NULL # removing the data (the app uses the data
         coming from other object)
94
95 # rename the lists and data.frame accordingly for the app
96 names(stats) <- c("df_predict","lt_predict","lt_summary",
```

```
 97                     "mu_plot")
 98 names(stats$df_predict) <- c("date", "q25",  "med",  "q975", "m")
 99 names(stats$lt_predict) <- c("date", "q25",  "med",  "q975", "m")
100 names(stats$lt_summary) <- c("NTC25","NTC500","NTC975",
101                             "high.dat.low","high.dat.med",
102                             "high.dat.upper","end.dat.low",
103                             "end.dat.med","end.dat.upper")
104
105 # prepare the list to be saved
106 list_out <- list(df_predict = stats$df_predict,
107                  lt_predict=stats$lt_predict,
108                  lt_summary=stats$lt_summary,
109                  mu_plot = stats$mu_plot)
110
111 i = 4 # for cases
112 # saveRDS - aggregated by the states
113 results_directory = "/home/CovidLP/Covid/app_COVID19/
        STpredictions/"
114 names(data_base$data) <- c("date","n","d","n_new","d_new")
115 file_id <- colnames(data_base$data)[i-2]
116 saveRDS(list_out,
117         file=paste0(results_directory,'Brazil_',file_id,'.rds'))
```

Listing 10.10: R code to aggregate the states outputs to build the results for Brazil.

Listing 10.10 collects the posterior prediction for each state and aggregates them. This is done in many steps: 1) a first state is used to initiate the variables (lines 11–28); 2) a loop on each state is started (line 31); 3) for each state, the predicted data is read and the variable (**data_base**) used to construct the Brazilian data set is updated (lines 34–75); 4) after the loop, the hidden variables necessary for the **pandemic_stats** function are created with the mean samples (lines 79–81); 5) with the aggregated data frame for Brazil created in the **data_base** variable, the summary statistics are created (line 92); and 6) the results are saved for the country (lines 96–117). It is important to notice that Listing 10.10 is executed necessarily after all states are fitted. This can be seen in Listing 10.8, where the **aggregate_brazil.R** script (Listing 10.10) is called in the sequence of the **brazil_predictions.R** script (Listing 10.7).

A perceptive reader must have noticed that although the **rstan** package was chosen to perform the Bayesian inference, it does not appear directly in any of the listings in this chapter. This is due to the fact that its use is hidden inside the functions of the **PandemicLP** package, which will be presented in details in Chapters 15 and 16.

A final necessary feature when conducting analysis is to check for the performance of the fitted models. This task was entrusted to the monitoring team, and discussion about it is presented in Chapter 13. Finally, after running all the scripts, the summary data of the fitted models are stored in the GitHub repository and will be accessed by the online application. In the next chapter, the reader can find details on how to construct an online application with the use of the **Shiny** package (Chang et al., 2020).

Bibliography

Analytics, R. and Weston, S. (2020) *doMC: Foreach Parallel Adaptor for 'parallel'*. URLhttps://CRAN.R-project.org/package=doMC. R package version 1.3.7.

Chang, W., Cheng, J., Allaire, J., Xie, Y. and McPherson, J. (2020) *shiny: Web Application Framework for R*. URLhttps://CRAN.R-project.org/package=shiny. R package version 1.5.0.

de Valpine, P., Paciorek, C., Turek, D., Michaud, N., Anderson-Bergman, C., Obermeyer, F., Wehrhahn Cortes, C., Rodrìguez, A., Temple Lang, D. and Paganin, S. (2020) *NIMBLE User Manual*. URLhttps://r-nimble.org. R package manual version 0.10.0.

Depaoli, S., Clifton, J. P. and Cobb, P. R. (2016) Just another Gibbs sampler (JAGS) flexible software for MCMC implementation. *Journal of Educational and Behavioral Statistics*, **41**, 628–649.

Gelman, A., Carlin, J. B., Stern, H. S., Dunson, D. B., Vehtari, A. and Rubin, D. B. (2013) *Bayesian Data Analysis*. New York: Chapman and Hall/CRC, 3rd edn.

Hoffman, M. D. and Gelman, A. (2014) The No-U-Turn sampler: Adaptively setting path lengths in Hamiltonian Monte Carlo. *Journal of Machine Learning Research*, **15**, 1351–1381.

Lunn, D., Spiegelhalter, D., Thomas, A. and Best., N. (2009) The BUGS project: Evolution, critique and future directions (with discussion). *Statistics in Medicine*, **28**, 3049–3082.

Lunn, D. J., Thomas, A., Best, N. and Spiegelhalter, D. (2000) WinBUGS – a Bayesian modelling framework: Concepts, structure, and extensibility. *Statistics and Computing*, **10**, 325–337.

Microsoft and Weston, S. (2020) *foreach: Provides Foreach Looping Construct*. URLhttps://CRAN.R-project.org/package=foreach. R package version 1.5.1.

Plummer, M. et al. (2003) JAGS: A program for analysis of Bayesian graphical models using Gibbs sampling. In *Proceedings of the 3rd International Workshop on Distributed Statistical Computing*, vol. 124, 1–10. Vienna, Austria.

Rue, H., Martino, S. and Chopin, N. (2009) Approximate Bayesian inference for latent Gaussian models by using integrated nested Laplace approximations. *Journal of the Royal Statistical Society: Series B (Statistical Methodology)*, **71**, 319–392.

Stan Development Team (2020) RStan: The R interface to Stan. URLhttp:
//mc-stan.org/. R package version 2.21.2.

11

Building an interactive app with Shiny

Thaís Paiva

Universidade Federal de Minas Gerais, Brazil

Douglas R. M. Azevedo

Localiza Rent a Car S.A., Brazil

Marcos O. Prates

Universidade Federal de Minas Gerais, Brazil

CONTENTS

The goal of this chapter is to offer a concise tutorial of how to build an online application to publish results of data analysis interactively. For this purpose, we relied mostly on the Shiny package (Chang et al., 2020) which offers a diverse set of functionalities. We also point to other sources for extending beyond the basic app features. This chapter covers good practices for app building and code organising, details on interactive plots, how to deploy and publish your app, and how to monitor access. All of these aspects are illustrated with examples from our experience in building the CovidLP application. This chapter will help first-time users, as well as those already familiar with Shiny, to build and improve their apps.

DOI: 10.1201/9781003148883-11

11.1   Getting started

The COVID-19 pandemic that affected the whole world in 2020 has brought the attention of the general public to updates and monitoring of time series of disease data, that otherwise were left mainly to specialists, such as statisticians and epidemiologists. This has added to many other motives behind the importance of publicising academic results and disseminating scientific knowledge. Because of this, many researchers have been seeking tools to publish data analysis in more dynamic forms than just academic papers. Examples of these tools are online dashboards and applications to display graphs and tables in an accessible and interactive way. The `Shiny` package (Chang et al., 2020) that creates interactive web applications has become popular for its versatility and not too steep learning curve. It is aimed at the public familiar with the R language, such as statisticians and data scientists, who are more interested in the data analysis process than the technical details of user interface programming.

This chapter is not intended to be a full extension tutorial of `Shiny` functions. For that, we leave the readers with the vast literature covering from the basics to the most advanced features of the package, such as Wickham (2021) and Colin Fay (2020). Our goal here is to cover the most important choices that were made during the development of the CovidLP app (available at `www.est.ufmg.br/covidlp`). With that in mind, we intend to provide a useful guide for the reader interested in building a similar application for displaying results of any kind of data analysis. Our guide is exemplified with epidemiological forecasts for the COVID-19 pandemic.

For a running example in the following sections, we will use the data series of new daily cases of COVID-19 from Argentina, from the beginning of the pandemic until November 15, 2020. This data set can be easily obtained with the `load_covid` function from the `PandemicLP` package that will be introduced later on Chapter 15. For an immediate replication of the running example, this specific data set is also available on our supplementary material online and can be loaded with the command in Listing 11.1.

```
data <- read.table(
  file = "https://raw.githubusercontent.com/CovidLP/Book/main/
    Part4/Chapter11/data/argentine.csv",
  sep = ";", header = TRUE
)
```

Listing 11.1: Getting data for the running example.

This `data` object has five columns: `date` is the current date, `cases` is the number of cumulative cases until the current date, `deaths` is the number of cumulative deaths until the current date, `new_cases` is the number of new cases reported on the current date, and `new_deaths` is the number of new deaths reported on the current date.

11.2 Shiny basics

The Shiny package was first released in 2012 aiming to help the R community to communicate results in a simple way. It is a package maintained by RStudio, an American company that, among other things, provides several R packages to the community. Since 2012, several new releases of Shiny have been presented, increasing even more the number of functionalities. Because of its flexibility and a shallow learning curve, it has attracted a huge number of new users and developers as well. Today it has around 15.5 million downloads showing its importance in our community. Due to its success, it is easy to find great extensions to the Shiny package that increase, even more, the number of functionalities and layout capabilities.

Before Shiny, building an interactive application was a headache for a statistician or data scientist, since it is not common to find such a professional with expertise in front-end programming. Therefore, the process of creating an app was slow and painful for such professionals. The Shiny package provides several built-in functions, tools, and default layouts that work as wrappers for HTML and CSS codes, and JavaScript functions. As a result, the developer can invest more time in their application and less time worried about front-end issues. However, the user is not limited to the built-in layouts, and knowledge in these other languages gives autonomy to the developer to create functionalities based on his necessities.

A Shiny app can be seen as an application of a function. Generally, a function receives some inputs/parameters, follows some instructions, and returns some outputs. When creating an app, the developer aims to provide an interface between the user and the code. The app interface contains several inputs (buttons) and outputs (plots, tables, text and so on). In this sense, the Shiny app will wait for the user's inputs, then execute some instructions to finally show the outputs.

Inputs and outputs are organised in a ui.R file, a short for "user interface" (this name is just a convention). The server.R file (another convention) has all the instructions to be evaluated as the inputs change. The files ui.R and server.R are related through reactivity. We can still have a global.R file containing constants, utility functions, and everything else that does not change interactively to be used throughout the code. The arrows in Figure 11.1 display the relationship between these three basic files.

In a Shiny app, all input and output elements are called by names (ids). These names must define uniquely each input or output. The input list can be accessed by evaluating input$id_input in the server.R file. Similarly, the outputs are created on the server side by assigning a result to an output position, let us say output$id_output. Just to make the logic understandable, we will present the code to create a simple app using the running example of Argentine data. The following app displays a time series of the cumulative number of cases of COVID-19 in Argentina until November 15, 2020.

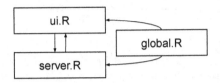

FIGURE 11.1: Relationship between standard `Shiny` files.

In the global file, we read the data from a remote source and in this case, it is static. We also transformed the date column to date format as presented in Listing 11.2.

```
data <- read.table(
  file = "https://raw.githubusercontent.com/CovidLP/Book/main/
    Part4/Chapter11/data/argentine.csv",
  header = TRUE
  )
data$date <- as.Date(data$date)
```
Listing 11.2: `global.R` file for our running example.

In the server file in Listing 11.3, we have a function of input and output containing one output element named `time_series` that is a scatter plot of a selected variable over time. The variable to be displayed will be selected by the user in the `ui.R` file in Listing 11.4. The available options are `cases`, `deaths`, `new_cases`, and `new_deaths`. The selected option is stored in the input element named `series`. The chosen `series` is displayed as a result of the `selectInput` function in Listing 11.4. Several types of inputs are available, such as `textInput`, `numericInput`, `dateInput` and `sliderInput`.

```
function(input, output){
  output$time_series <- renderPlot({
    series <- input$series
    plot(x = data[["date"]], y = data[[series]],
         xlab = "Date", ylab = series)
  })
}
```
Listing 11.3: `server.R` file for our running example.

```
fluidPage(
  selectInput(
    inputId = "series",
    label = "Select a variable",
    choices = c("cases", "deaths", "new_cases", "new_deaths"),
    selected = "cases"
  ),
  plotOutput(outputId = "time_series")
)
```
Listing 11.4: `ui.R` file for our running example.

To interact with the Shiny app locally, just run the function runApp(appDir="app_dir") replacing "app_dir" by the path to your application folder. In this case, the Shiny app will be visible just for you. To share it in a network or even online, see Section 11.7.

As mentioned before, all of these functions are wrappers for HTML codes and JavaScript functions. For example, the output code of the selectInput function is given in Listing 11.5. In this sense, our work is facilitated since we do not need to code in HTML to create a button.

```
1  <div class="form-group shiny-input-container">
2    <label class="control-label" for="series">Select a variable</
       label>
3    <div>
4      <select id="series"><option value="cases" selected>cases</
         option>
5  <option value="deaths">deaths</option>
6  <option value="new_cases">new_cases</option>
7  <option value="new_deaths">new_deaths</option></select>
8      <script type="application/json" data-for="series" data-
         nonempty="">{}</script>
9    </div>
10 </div>
```

Listing 11.5: HTML output generated by the selectInput function.

The time series plot is displayed in the app by accessing its name through the plotOutput function in the ui.R file. In the same way, the plot is created in the server side (server.R) by using the function renderPlot. For each kind of output, we have different *Output and render* functions. Table 11.1 shows a list containing some of them.

TABLE 11.1: Some important output functions.

server side	ui side	description
renderPlot()	plotOutput()	Basic R plots
renderTable()	tableOutput()	Table
renderText()	textOutput()	Text
renderImage()	imageOutput()	images
renderDataTable()	dataTableOutput()	Interactive tables
renderPlotly()	plotlyOutput()	Plotly graphs

The last ingredient in our app is the fluidPage function that basically wraps everything on a single page. To have more control about the layout, one can explore some built-in functions as the navbarPage, navlistPanel, and sidebarlayout. Figure 11.2 shows the resulting app produced with the files above.

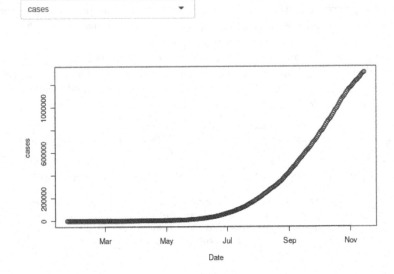

FIGURE 11.2: Screenshot of the resulting `Shiny` app for the running example.

As aforementioned, the key concept of a `Shiny` app is the reactivity that happens when the user interacts with the app contents. Therefore, the `server.R` file must be reactive to users' choices. There are two key concepts for understanding the reactivity: 1) it should react fast as the user interacts with the app; 2) it does not need to execute the entire `server.R` file every time, but just the elements that relate to what the user chooses. The latter is known as laziness and becomes important as the app increases in complexity. For example, if your app contains three independent outputs –so you have at least three entries in the `server.R` file– and the user is interacting with inputs that only affect one of them, it is not necessary to update all the three outputs, but just one. It is particularly important for your `Shiny` app to have good performance, because it can avoid the execution of unimportant pieces of the script saving time and displaying results to your users as fast as possible.

An implication of the laziness property is that the execution order of the `server.R` file is not the customary top to bottom. It can be evaluated in different orders according to the complexity of the app and the user's choices. In our previous example, the `output$time_series` depends only on the variable selected in `input$series`, so we say that there is a reactive dependency between `output$time_series` and `input$series`, and the `output$time_series` is updated only if a different `input$series` is selected.

There are several functionalities in the `Shiny` package that allow us to manage reactivity. The `reactive` function creates a reactive expression allowing the developer to execute some slow computation just when it is needed. In our example, we can create a reactive expression that selects the column we want to display. Also, we can add a `dateInput` to control the start date of the time series to display. If the changes in the start date are not related to those in the variable selection, then our reactive function should only be activated with changes in `input$series`. Listing 11.6 shows how our `server.R` would look in such a situation, and 11.7 shows how to add the `dateInput` in our `ui.R` file.

```
function(input, output) {
  reactive_data <- reactive({
    series <- input$series
    ts_plot <- data[, c("date", input$series)]

    return(ts_plot)
  })

  output$time_series <- renderPlot({
    series <- input$series
    start_date <- input$start_date

    data <- reactive_data()
    data_sub <- subset(data, date >= start_date)

    plot(x = data_sub[["date"]], y = data_sub[[series]],
         xlab = "Date", ylab = series
    )
  })
}
```

Listing 11.6: Using the `reactive` function in our app to select the correct column to display.

```
fluidPage(
  selectInput(
    inputId = "series",
    label = "Select a variable",
    choices = c("cases", "deaths", "new_cases", "new_deaths"),
    selected = "cases"
  ),
  dateInput(
    inputId = "start_date",
    label = "Start date",
    value = "2020-01-23",
    min = "2020-01-23",
    max = "2020-11-15"
  ),
  plotOutput(outputId = "time_series")
)
```

Listing 11.7: Using the `dateInput` function to select the first date to display.

The reactive expression will be evaluated if an input of the expression changes (it can be more than one, but in this case, is just `input$series`). When the user selects a different start date, just the output is going to change, which means that the reactive expression is not called in this case. The reactive expression returns a dataset with two columns that can be evaluated in the `renderPlot` via `reactive_data()`.

Similarly, the `Shiny` package has the `eventReactive` function that also works as a reactive expression, but the developer can choose which events should call the reactivity. The functions `observe` and `observeEvent` also provide similar functionalities, but instead of returning outputs to be used on the server side, it can be employed to update widgets on the ui side. Another useful function is the `isolate` function, which isolates the reactivity of a desired input on the server side.

11.3 Beyond `Shiny` basics

We have illustrated some nice basic functionalities of the `Shiny` package. However, `Shiny` is even more powerful and provides several advanced tools and extensions for those who want to build more complex apps (Wickham, 2021).

Although `Shiny` has several built-in functionalities, it might be challenging to create a nice-looking `Shiny` app using only the basic commands. As the number of users have increased in the past 10 years, the number of `Shiny` extensions also did, providing awesome add-ons to the basic functionalities. In this section, we provide some references for those who want to elevate the level of their `Shiny` apps.

It is expected that an appealing app has a good and clear presentation of its content. Therefore, the visual aspect of a `Shiny` app is very important to its success. Several pre-built layouts are available as `Shiny` extensions, as the case of the `shinydashboard` package (Chang and Borges Ribeiro, 2018). It facilitates the creation of dashboards, providing several visual improvements for the basic `Shiny` layouts. Among several great options, we can cite the `shinymaterial` (Anderson, 2020) and `argonR` (Granjon, 2019) packages for creating nice-looking apps.

We all agree that `Shiny` basics provide several useful widgets for user interaction. However, the package `shinyWidgets` (Perrier et al., 2020) extends the widgets to the next level. The best characteristic of this package is that the functions and argument names are similar to those from the `Shiny` package, which makes it easy to adapt the code.

A self-contained application needs to dialogue with the user by exposing the ideas behind the inputs and explaining some of the outputs. This communication can be done via alerts, popups, tooltips, modals, and other interactive contents. Again, several extensions are available, and among them we recommend the package `shinyBS` (Bailey, 2015) for nice-looking alerts, popups, tooltips and modals.

While the application is performing tasks that take longer, it is good to show the user an indication that the work is being done. A nice package for this task is the `shinycssloaders` (Sali and Attali, 2020), which adds loading animations to a `Shiny` output while it is recalculating.

Sometimes you need to hide/show, disable/enable some inputs. These are common JavaScript operations which you can find in the `shinyjs` package (Attali, 2020). Also, for those that have some knowledge in JavaScript, this package facilitates the integration of JavaScript functionalities.

Another nice extension is the `golem` package, which is a toolkit that simplifies the creation, development, and deployment of a `Shiny` application. Using the `golem` package from the very beginning can avoid problems as the app size increases and the changes are being incorporated.

For a full detailed list of `Shiny` extensions we recommend visiting `github.com/grabear/awesome-rshiny`.

11.4   Code organisation

An important aspect that is not directly related to the functions of a `Shiny` app, but also critical to the development of most projects, is the code organisation. Even though one can create an app without taking care of code organisation, following some good organising practices can optimise maintenance time and facilitate the code flow understanding. Code organisation is not related to the instructions given to the computer, but how easy it is for developers to understand the code.

The organisation of a `Shiny` app is particularly important in three situations: 1) the app has several functionalities and hence has a huge number of code lines; 2) the app is complex and depends on the integration of several different programming languages; and 3) several developers are working in the same app. In these cases, it is easy to waste several hours searching for a small bug in a disorganised code flow. For these reasons, we provide in this section some tips on how to organise the code files of your app according to our experience on the CovidLP project.

We can divide the code organisation into best practices for coding and for project organisation. For the former, a good reference is the *"The Tidyverse Style Guide"* (`style.tidyverse.org`) in which you will find several tips for writing understandable code. The tips can differ from one reference to another, however, the most important is to define some rules to be employed consistently by the entire team. For the latter, the structure of the files will depend on the project's necessities. However, it is possible to start with a default `Shiny` structure that is applied in several cases, including our app, as can be seen in the directory tree in Figure 11.3. As mentioned, this structure can be adapted for each project according to its necessities.

We have created three folders under our working directory. The `html` folder has files in HTML format that are used by the `ui.R` file, as is the case of the

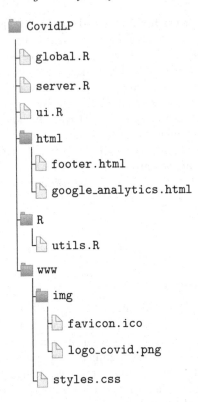

FIGURE 11.3: Directory tree of the CovidLP app.

footer of our app and the Google Analytics code used to monitor the access to our app. In the R folder, we have some functions to download data and create plots that are used by both server.R and ui.R files. Organising these functions in a separate file is important to avoid code duplication as we use these pieces of code several times in our app.

The www folder has style customisation options in the styles.css file, and also some images in the img folder. The www folder is a convention and it is expected to store the elements that will be presented in the users' screen, such as layouts (.css files), images, videos and so on.

11.5   Design of the user interface

There is a large set of functions to help us create the user interface that can be the difference in a nice interactive app. The list of layouts is huge and they cover several conventional website layouts as fluid pages, pages with a sidebar, dashboards, and much more.

For the sake of illustration, in Listing 11.8 we change the layout of our running example to have a sidebar for the input buttons. Inside the `fluidPage` element, we added the `sidebarLayout` function to create an app with widgets in the sidebar and a main panel containing our outputs.

```
fluidPage(
  sidebarLayout(
    sidebarPanel = sidebarPanel(
      width = 2,
      selectInput(
        inputId = "series",
        label = "Select a variable",
        choices = c("cases","deaths","new_cases","new_deaths"),
        selected = "cases"
      ),
      dateInput(
        inputId = "start_date",
        label = "Start date",
        value = "2020-01-23",
        min = "2020-01-23",
        max = "2020-11-15"
      )
    ),
    mainPanel = mainPanel(
      plotOutput(outputId = "time_series")
    )
  )
)
```

Listing 11.8: Using the `sidebarLayout` function to create an app with a sidebar and a main area.

The `sidebarLayout` function wraps all the content of our app, and the `sidebarPanel` and `mainPanel` functions inform which elements go to the sidebar and to the main panel, respectively. The `width` parameter on line 4 of Listing 11.8 controls the size of the sidebar panel, and it should be a number between 1 and 12. Therefore, we are reserving a slot of width 2 for the widgets, and hence a slot of width 10 will be left for the main panel plot. The resulting app with the sidebar panel is displayed in Figure 11.4.

What if we wanted the buttons side by side? In this case we could use the useful `column` function. It helps organise inputs and outputs in the user interface. The `column` has a parameter `width` that is an integer between 1 and 12. The number of possible columns is 12 because 12 is divisible by 1, 2, 3, 4, 6 and 12, which provides several ways to equally divide the screen. However, it is possible to use the `column` function sequentially dividing the screen even more. For example, `column(width = 1, column(width = 5, ...))` will produce a column with size $5/(12*12)$ of the screen size.

In Listing 11.9 we added the `fluidRow` element function wrapping two `column` elements, each occupying half of the available width of the side panel. The `fluidRow` helps with the delimitation of the content in a line, and in

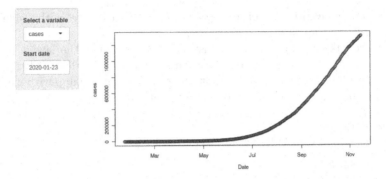

FIGURE 11.4: Screenshot of the resulting **Shiny** app using the **sidebarLayout** function.

this sense it works similarly to the **column** function. Figure 11.5 presents the resulting app interface.

```
1  fluidPage(
2    sidebarLayout(
3      sidebarPanel = sidebarPanel(
4        width = 2,
5        fluidRow(
6          column(
7            width = 6,
8            selectInput(
9              inputId = "series",
10             label = "Select a variable",
11             choices=c("cases","deaths","new_cases","new_deaths"),
12             selected = "cases"
13           )
14         ),
15         column(
16           width = 6,
17           dateInput(
18             inputId = "start_date",
19             label = "Start date",
20             value = "2020-01-23",
21             min = "2020-01-23",
22             max = "2020-11-15"
23           )
24         )
25       )
26     ),
27     mainPanel = mainPanel(
28       plotOutput(outputId = "time_series")
29     )
30   )
31 )
```

Listing 11.9: Using the **fluidRow** and **column** functions to divide the interface area.

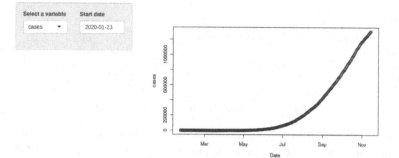

FIGURE 11.5: Screenshot of the resulting Shiny app using the column and
fluidRow functions.

Sometimes we want some design customisation that might not be covered
by the Shiny basic capabilities. It is common to get stuck trying to change
widgets' positions and layout colours. To follow a specific pattern, we can use
a css file to control these types of settings. For example, suppose that we
want to change the colour of the sidebar panel, instead of the default grey.
For that, we have at least two easy alternatives. The first one is adding the
parameter style = 'background-color: grey' with the desired colour to
the sidebarPanel function, as in line 5 of Listing 11.10.

```
fluidPage(
  sidebarLayout(
    sidebarPanel = sidebarPanel(
      width = 2,
      style = 'background-color: grey', # Here!
      fluidRow(
        column(
          width = 6,
          selectInput(
            inputId = "series",
            label = "Select a variable",
            choices=c("cases","deaths","new_cases","new_deaths"),
            selected = "cases"
          )
        ),
        column(
          width = 6,
          dateInput(
            inputId = "start_date",
            label = "Start date",
            value = "2020-01-23",
            min = "2020-01-23",
            max = "2020-11-15"
          )
        )
      )
    ),
    mainPanel = mainPanel(
```

```
29      column(
30        width = 6,
31        plotOutput(outputId = "time_series")
32      )
33    )
34  )
35 )
```

Listing 11.10: Using the style property to design the user interface.

A second alternative is to include a `css` file that can contain several layout instructions for the `Shiny` app. This is done using the `includeCSS` function and passing the `css` file location as a parameter anywhere in the `ui.R` file, as in line 2 of Listing 11.11. Listing 11.12 presents the content of the `www/styles.css` file setting the parameter for the background colour. The main advantage of using a `css` file is that one can create a layout just once, and apply it to all widgets that will follow the same pattern, avoiding the `style` parameter. In both alternatives, the resulting app, with the different sidebar background colour, can be seen in the screenshot of Figure 11.6.

```
1  fluidPage(
2    includeCSS("www/styles.css"), # Here!
3    sidebarLayout(
4      sidebarPanel = sidebarPanel(
5        width = 2,
6        fluidRow(
7          column(
8            width = 6,
9            selectInput(
10             inputId = "series",
11             label = "Select a variable",
12             choices=c("cases","deaths","new_cases","new_deaths"),
13             selected = "cases"
14           )
15         ),
16         column(
17           width = 6,
18           dateInput(
19             inputId = "start_date",
20             label = "Start date",
21             value = "2020-01-23",
22             min = "2020-01-23",
23             max = "2020-11-15"
24           )
25         )
26       )
27     ),
28     mainPanel = mainPanel(
29       plotOutput(outputId = "time_series")
30     )
31   )
32 )
```

Listing 11.11: Including a `css` file to design the user interface.

```
1  .well{
2      background-color: grey;
3  }
```

Listing 11.12: css file to change some style properties of the user interface.

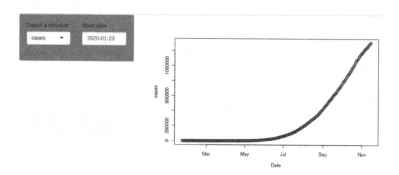

FIGURE 11.6: Screenshot of the resulting `Shiny` app by adding a simple `.css` file.

11.5.1 The CovidLP app structure

After introducing the different elements that can be used to create a user interface for a `Shiny` app, we will describe now the functions that were used to create the current version of the CovidLP online application. These elements are defined in the `ui.R` file, following the `Shiny` structure that was described in Section 11.2. All the source code of the app, including the `ui.R` file, can be accessed on our GitHub repository (`github.com/CovidLP/App`).

The app main page is organised within a `FluidPage` component. This page, in turn, is further divided into `FluidRow` elements. The first row is the page header, and contains the page title and some important links, organised in columns. The second row contains the information about how to interact with the app by selecting the country/state to see the data, inserted inside a `wellPanel` element with an inset border and grey background. All of these elements are styled according to the parameters specified in the `styles.css` file as mentioned in Section 11.5. A screenshot of these first elements can be seen in Figure 11.7.

After that, we have a highlighted bar that holds the input buttons for selection of country and state, also inside a `FluidRow` element. The input buttons are included with a `pickerInput` function from the `shinyWidgets` package, which provides extra functionalities for the `Shiny` apps discussed in Section 11.3. On this input bar, we also included a `tipify` button from the `shinyBS` package, which displays an information message when the user hovers over the button. These elements are shown in Figure 11.8.

FIGURE 11.7: Screenshot of the first elements of the user interface of the CovidLP app.

FIGURE 11.8: Screenshot of the input bar of the user interface of the CovidLP app.

The fourth `FluidRow` holds the tabs and plots with the prediction results. This row contains a `tabsetPanel` with three tabs for display of the data, short-term predictions, and long-term predictions. Each tab has a region on top for the user to select the kind of data to display (confirmed cases and/or deaths) and the option to use a logarithmic scale for the y-axis of the plot. These inputs are added using HTML tags, and with the `awesomeCheckboxGroup` `awesomeCheckbox` functions from the `shinyWidgets` package. Here, we also added a `bsTooltip` element from the `shinyBS` package to display a help message when the user hovers over the selected region. The screenshot in detail of this part is shown in Figure 11.9.

FIGURE 11.9: Screenshot of the three tabs of the user interface of the CovidLP app.

Finally, the plots are added with the `plotlyOutput()` function that prepares a `plotly` object, as it will be described in Section 11.6, to be an output element of the app. These plot functions are wrapped in a `withSpinner()` function from the `shinycssloaders` package, that displays a spinner or other selected animation that indicates when an output is being refreshed. We include a screenshot of the app when the first tab is selected in Figure 11.10, but for the sake of the page size, only the first plot of the time series of new cases in Argentina is shown.

FIGURE 11.10: Screenshot of the data tab of the user interface of the CovidLP app, with Argentina as the selected country.

After the plots, we also included some download buttons, with the `downloadBttn()` from the `shinyWidgets` package, to allow the user to download the data that was used to make the plots. These data files are prepared in a separate reactive function, since it depends on the input selected. Finally, after the plots and the download buttons, the last element of the app is the footer of the page, included as an HTML code. The screenshot of these last parts is in Figure 11.11.

FIGURE 11.11: Screenshot of the bottom elements of the user interface of the CovidLP app.

Another extra interface element that we added to our app, after noticing that a large part of the audience accessed the initial version of the app on mobile, is a pop-up window when the page first loads. This window shows a message informing that the app is best viewed on mobile with the device on horizontal. This element is actually included on the `server.R` file, with the function `showModal()` function. A screenshot of the pop-up window is shown in Figure 11.12.

We want to point out that there are an infinity of options and functions to design and style an app's user interface. Here, we just explained our choices in the current version of the CovidLP app as an example for the reader.

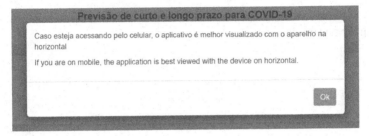

FIGURE 11.12: Screenshot of the pop-up window that is shown on start of the CovidLP app.

11.6 Creating interactive plots

This section provides a brief guideline on how to create interactive plots. Section 11.6.1 presents an introduction to the `plotly` package (Sievert, 2020), and Section 11.6.2 describes the main steps of the process of designing the interactive plots shown in our online application.

First of all, we decided that it was necessary to allow some sort of user interaction with the plots, since there are many different data perspectives that could be investigated by the users. For example, it might be useful to show the data values when hovering over the points of a large data series, or zooming into specific areas of the plot to take a closer look at what is going on locally. These and more are some of the functionalities of the plot outputs produced with the `plotly` package. These attributes were also a good fit to our intention to publish frequently updated results on a user-friendly web-based application, as it was described in the previous sections.

Here we will describe the graphical elements of the basic plots, and focus on the specific functions that were used in our application of the CovidLP project. For a more comprehensive reference, the book *"Interactive Web-Based Data Visualization with R, plotly, and Shiny"*, available on `plotly-r.com`, by Sievert (2020) provides a detailed tutorial of how to build all sorts of interactive graphs with the help of the `plotly` and `Shiny` packages.

11.6.1 `plotly` basics

After loading the `plotly` package, any interactive plot can be initialised with the function `plot_ly()`, that should receive as its first argument the data frame with the set of points to be plotted. This can be done using the pipe operator `%>%`, which feeds the object on its left side as the first argument of the function on the right. As a simple example, Listing 11.13 shows how to create an object `plt` that will hold the plot, which is empty for now. We continue to use the data of new daily cases from Argentina as our running example. We will assume that the data is already loaded in the `data` object, with columns for the series of dates and new cases.

```
1  plt = data %>%
2      plot_ly()
```

Listing 11.13: Initialising a `plotly` object.

To actually draw the plot elements, it is necessary to add a trace component that can be of type scatter, histogram, pie, etc. This can be done with the `add_trace` function which is part of the `add_*` family (that includes the functions `add_lines` and `add_histogram`) to render data into geometric objects. These objects can be seen as layers of the plot, and should include the elements of the data, aesthetic mappings, geometric representation, and other transformations.

In our example, we select the scatter plot type, and set the plot mode to connect the points with lines, with the commands in line 3 of Listing 11.14. After that, the plot can be seen by calling the `plt` object which will display the graph shown in Figure 11.13. For the exhaustive list of all the options of the `plotly` functions, visit `plotly.com/r/reference/index`.

```
1  plt = plt %>%
2      add_trace(x=data$date, y=data$new_cases,
3              type='scatter', mode='lines+markers',
4              name="new daily cases",
5              marker=list(color="black"), line=list(color="grey")
       )
```

Listing 11.14: Adding a trace element to the `plotly` object.

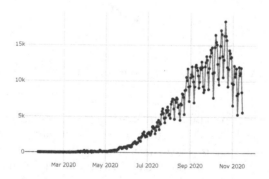

FIGURE 11.13: Basic `plotly` graph generated with the code in Listings 11.13 and 11.14.

Next, the user can set several plot layout options with the function `layout()`. This function takes as the first argument a `plotly` object, such as the object `plt` that we created in the previous listing, and this can also be done using the pipe operator. For example, if we want to add a plot title, axes labels and a legend to the graph created before, we can continue adding to the `plt` object and execute in sequence the commands in Listing 11.15. The resulting plot, after setting the layout options, can be seen in Figure 11.14. Note that we added a title with the country name, axes labels with the variable names, and a legend with a grey background positioned on the upper left corner. The arguments of the `layout()` function are lists that contain several options for each plot element. For the reference of the format of all the arguments of this function, access `plotly.com/r/reference/layout`.

```
1  # add title, axes labels and legend
2  plt = plt %>%
3     layout(title=list(text="Argentina"),
4             xaxis=list(title="Date"),
5             yaxis=list(title="number of cases"),
6             legend=list(x=0.1,y=0.9, bgcolor=gray(0.9)),
7             showlegend=T
8             )
9  plt
```

Listing 11.15: Configuring the layout of a `plotly` object.

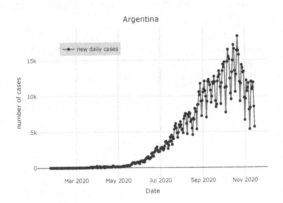

FIGURE 11.14: `plotly` graph with some layout options generated after running the code in Listing 11.15.

Lastly, we want to highlight the `config` function of the `plotly` package. This function can be used to configure some elements of the graph such as the modebar (bar that is revealed on the top of the plot with some action buttons for the graph), file download formats, language support and so on. For more details about this function, check the `config` topic of the `plotly` documentation on `rdocumentation.org/packages/plotly`.

11.6.2   The CovidLP plots

Following the same track of the previous section, we describe now the structure of the interactive plots shown in our online application (available at `est.ufmg.br/covidlp`). The basic structure of all of our plots can be seen in

Listing 11.16. Again, we create the object `plt` which contains all the elements of the plot, and when it is called on its own, such as in line 7, it will display the current graph. As before, the first element of the `plot_ly()` function is the data frame with the values to be plotted, here named `data`. In this case, it is the output of a reactive function that returns the appropriate values depending on the input that the user selects. For more details about the reactive functions, refer back to Section 11.2.

```
1  plt <- data %>%
2      plot_ly() %>% # main call of the plotly function
3      config( "config args" ) %>%
4      layout ( "layout args" ) %>%
5      add_annotations( "args" ) %>%
6      add_trace( "args" )
7  plt
```

Listing 11.16: Basic structure of the `plotly` graphs from our app.

After the main call of the `plot_ly()` function on line 2 of Listing 11.16, there are some varying functions that will depend on the specific characteristics of each graph.

Some of the configurations that we used are related to the `plotly` modebar that adds several buttons to the top of the plot. These buttons allow the user to zoom in and out, reset the view, download the plot, and many more actions. For more details on how to configure the modebar, see Chapter 26 of Sievert (2020). Inside the `config` function, it is also possible to select the file download format and customise the file name for a more informative version.

All the added text, such as the source reference and updated date, is added as annotations via the function in line 5. The function `add_annotations` is used to add text elements to any position of the plot.

Lastly, but most importantly, the main elements of the plots, such as data points and prediction points and intervals, are added as traces with the function of line 6. This function is repeated for each of the different elements that we include in our graphs, each of them with its specific set of arguments. All of the source code used to produce our app's plots is publicly available in our GitHub repository (`github.com/CovidLP/App`).

11.7   Deploy and publish

You have finally finished your **Shiny** app and now it is time to share with the world. Handing out a project with several R files may not attend the purpose of creating an interactive app. This would require knowledge to understand the app structure and minimal knowledge of the code by the users.

In this case, the developer can invest time in three main directions for publishing the app online. The first and easier alternative is to use servers already prepared for hosting `Shiny` applications. That is the case of the `shinyapps.io`, a RStudio product. Because of that, it is really easy to deploy an app either from the RStudio graphical interface or using some simple commands shown in Listing 11.17.

```
library(rsconnect)
setAccountInfo(name='<ACCOUNT NAME>',
               token='<TOKEN>',
               secret='<SECRET>')
deployApp(appDir = "path/to/app", appName = "myapp_name")
```

Listing 11.17: Deploying an app in shinyapps.io.

The fields `<ACCOUNT NAME>`, `<TOKEN>` and `<SECRET>` are keys provided by `shinyapps.io` after creating an account. To deploy an app to your account, you just need to set up your account info and then upload the app using the command `deployApp`. After that, your application will be available at the URL `account_name.shinyapps.io/myapp_name`. A nice thing about `shinyapps.io` is that you will have a small machine available for free so it is possible to host some apps without any cost.

Another good option is `digitalocean.com`, and for that, several tutorials are available on the internet showing how to set up and deploy a `Shiny` app, see for example Attali (2015). However, in this case it is necessary to pay for the use of a machine even for small applications.

For those who are working with sensitive information or already have a private server, it is possible to set up a `Shiny` server on it. For Linux users, basically you will need to install the `R` software, the packages needed, and the `Shiny server` (also a RStudio product `rstudio.com/products/shiny/shiny-server`) on the machine. After that, a port will be available (3838 by default) and a folder `shiny-server` will be responsible to store the apps (on the `shiny-server` local installation). Finally, you can move your `myapp_name` folder with all source files to the `shiny-server` folder, and then your application will be available at `localhost:3838/myapp_name`. For a full tutorial on how to deploy on Linux servers, see `shiny.rstudio.com/articles/shiny-server.html`. For Windows and MAC users, the solution is similar, however a Linux virtual machine or a container will be necessary before the first step.

Another good option is to create a standalone application. In this case, the package `eletricShine` (Clark, 2020) is a good option. A downside is that it is currently only available for Windows systems. The package `golem` (Guyader et al., 2020) provides some useful functions for those who want to deploy an app in `engineering-shiny.org/deploy-golem.html`.

11.8   Monitoring usage

After publishing an app online, one may be curious about the traffic in the page. The easiest way to track the usage of a `Shiny` app is the Google Analytics solution. It will work independently where you published your app. A nice tutorial can be found at `shiny.rstudio.com/articles/google-analytics.html`.

The first step is to create an account on Google Analytics by accessing `marketingplatform.google.com/about/analytics`. It is possible to create a free account with several functionalities, but there are also paid options with more features.

Inside your account, you must to create a property (that is your app). In this step you must provide a web address to inform Google which address to track. After these two simple steps, the website will provide a tracking ID number and a JavaScript script. It should look like the code in Listing 11.18 in which **UA-9999999999-9** represents your tracking ID.

```
<!-- Global site tag (gtag.js) - Google Analytics -->
<script async src="https://www.googletagmanager.com/gtag/js?id=UA
    -160639476-1"></script>
<script>
  window.dataLayer = window.dataLayer || [];
  function gtag(){dataLayer.push(arguments);}
  gtag('js', new Date());

  gtag('config', 'UA-9999999999-9');
</script>
```

Listing 11.18: Deploying an app in shinyapps.io.

Saving this code inside a `file_name.html` file enables us to incorporate the tracking in our app. To do so, we can simply insert `tags$head(includeHTML(("file_name.html")))` anywhere in the `ui.R` file and that is all. Now you are able to follow the traffic in your **Shiny** app by accessing your Google Analytics account. For example, Figure 11.15 shows some traffic statistics in a 7-day window of the CovidLP app.

For those using `shinyapps.io` to host the app, it is also possible to have some statistics accessing your account. Although it is easy to track users

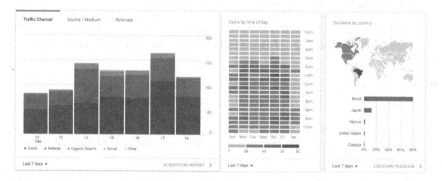

FIGURE 11.15: Some statistics of the CovidLP traffic in a 7-day window provided by Google Analytics.

with this approach, it has fewer visualisation options than Google Analytics. Figure 11.16 shows some traffic statistics of the CovidLP app from the `shinyapps.io` website.

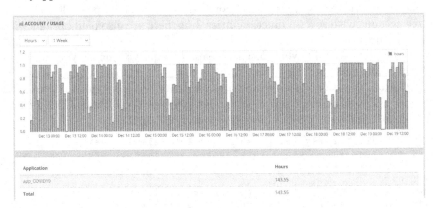

FIGURE 11.16: Some statistics of the CovidLP traffic in a 7-day window provided by `shinyapps.io`.

Bibliography

Anderson, E. (2020) *shinymaterial: Implement Material Design in Shiny Applications.* URLhttps://ericrayanderson.github.io/shinymaterial/. R package version 1.2.0.

Attali, D. (2015) How to get your very own RStudio Server and Shiny Server with DigitalOcean. Website. URLhttps://deanattali.com/2015/05/09/setup-rstudio-shiny-server-digital-ocean/.

— (2020) *shinyjs: Easily Improve the User Experience of Your Shiny Apps in Seconds.* URLhttps://deanattali.com/shinyjs/. R package version 2.0.0.

Bailey, E. (2015) *shinyBS: Twitter Bootstrap Components for Shiny.* URLhttps://ebailey78.github.io/shinyBS. R package version 0.61.

Chang, W. and Borges Ribeiro, B. (2018) *shinydashboard: Create Dashboards with 'Shiny'.* URLhttp://rstudio.github.io/shinydashboard/. R package version 0.7.1.

Chang, W., Cheng, J., Allaire, J., Xie, Y. and McPherson, J. (2020) *shiny: Web Application Framework for R.* URLhttps://CRAN.R-project.org/package=shiny. R package version 1.5.0.

Clark, C. (2020) *electricShine: Create Distributable Shiny Electron Apps.* URLhttps://github.com/chasemc/electricShine. R package version 0.0.0.9000.

Colin Fay, Sébastien Rochette, V. G. C. G. (2020) *Engineering Production-Grade Shiny Apps.* URLhttps://engineering-shiny.org/.

Granjon, D. (2019) *argonR: R Interface to Argon HTML Design.* URLhttps://github.com/RinteRface/argonR. R package version 0.2.0.

Guyader, V., Fay, C., Rochette, S. and Girard, C. (2020) *golem: A Framework for Robust Shiny Applications.* URLhttps://CRAN.R-project.org/package=golem. R package version 0.2.1.

Perrier, V., Meyer, F. and Granjon, D. (2020) *shinyWidgets: Custom Inputs Widgets for Shiny.* URLhttps://github.com/dreamRs/shinyWidgets. R package version 0.5.4.

Sali, A. and Attali, D. (2020) *shinycssloaders: Add Loading Animations to a 'shiny' Output While It's Recalculating.* URLhttps://github.com/daattali/shinycssloaders. R package version 1.0.0.

Sievert, C. (2020) *Interactive Web-Based Data Visualization with R, plotly, and shiny.* Chapman and Hall/CRC. URLhttps://plotly-r.com.

Wickham, H. (2021) *Mastering Shiny: Build Interactive Apps, Reports, and Dashboards Powered by R.* O'Reilly Media, Incorporated. URLhttps://mastering-shiny.org/index.html.

Part V

Monitoring

12

Daily evaluation of the updated data

Vinícius D. Mayrink
Universidade Federal de Minas Gerais, Brazil

Juliana Freitas
Universidade Federal de Minas Gerais, Brazil

Ana Julia A. Câmara
Universidade Federal de Minas Gerais, Brazil

Gabriel O. Assunção
Universidade Federal de Minas Gerais, Brazil

Jonathan S. Matias
Universidade Federal de Minas Gerais, Brazil

CONTENTS

This chapter can be seen as an overview describing the main difficulties and challenges related to the analysis of the pandemic data collected in a daily basis. Monitoring the behaviour of the new observations included in the epidemiological time series, for each region, is a central task to explain and anticipate a poor model fit that may lead to incorrect inferences. Modelling adaptations are required to handle the detected issues, therefore, the daily visualisation of the updated series is an important step to improve the statistical analysis throughout the study. The purpose of the present chapter is to describe the main scenarios of atypical counts or trends detected in one or more time series representing different regions. This topic is addressed in the context of the COVID-19 outbreak, but the reader should bear in mind that the obstacles reported here may also be found in other large epidemiological studies.

DOI: 10.1201/9781003148883-12

12.1   The importance of monitoring the data

The CovidLP project is a broad study involving a large number of regions. Particular attention is devoted to investigate the progress of the COVID-19 in each Brazilian state. In this case, the detailed Brazilian data were obtained from the daily reports released by the country's Ministry of Health, which collects the information from the local health authorities in each state. The daily data by notification date from government official sources of other countries were conveniently compiled and saved as time series summary tables available from a repository managed by the Center for Systems Science and Engineering (CSSE) at Johns Hopkins University; see https://coronavirus.jhu.edu/map.html and Dong et al. (2020). An important aspect to be highlighted about the data is the fact that the government policy regarding collection, processing and release of the pandemic information may differ among the regions under study. This point can be used to justify the occurrence of some specific features in the trajectory of the time series for a given country. A few examples of such features are: (*i*) the time delay between detection and reporting of an infected case or death may determine inflated counts for some points in the time series, (*ii*) lower counts may be observed on weekends or holidays since few professionals are working on processing the information, (*iii*) presence of atypical observations due to recounting procedures to correct inconsistencies or false positive identifications, which can be a part of the data management routine, and (*iv*) magnitude alterations related to changes in the format of the data being released. Data management is not the only explanation for distinctions between the series from different countries. In fact, the way in which the pandemic is fought by the authorities and the population affects the progress of the epidemiological curve; for example, a plateau or two/three growth waves could be observed in some cases.

The Brazilian Ministry of Health[1] (hereafter BMH in this chapter) is the government sector responsible for the administration and maintenance of policies related to public health in the country. During the COVID-19 outbreak, the BMH was in charge of collecting, preparing and releasing the national official count data indicating the progress of the pandemic. This task is extremely convenient for scientific studies, since it provides in a single website all sequential data sets per region and for the whole country. The information is also important for the press, which was interested in summarising through descriptive statistics the current scenario for the general audience watching or reading the daily news. Despite this importance, there were some specific issues caused by technical alterations and decisions adopted by the BMH on how to organise the official data for the public. This represented a major challenge to preserve the release of daily updated projections. In some moments, the

---

[1] https://www.gov.br/saude/pt-br

BMH electronic address to access the data files was changed, which forced an adaptation in the computational routine to download the data. This address change could be detected through inspection, done via the daily monitoring, after verifying that the newest observations were not updated to generate the current predictions. Another critical issue causing delays in the data analyses was the BMH decision to modify the original file type (CSV) to a different format (PDF) more complicated to read in statistical softwares. Fortunately, this alteration was undone a few days later. Perhaps the most problematic issue occurred in July 2020 when the BMH decided to report only the number of deaths and cases observed in the last 24 hours, excluding here (which was not previously done) the past cases that could not be processed in the same day of occurrence. This aspect determined a dramatic scale change for the count observations, leading to an unusual time series that could not be evaluated by the statistical model adopted for the whole pandemic period. In summary, the monitoring played a key role in the CovidLP project to identify issues associated with the lack of consistency in the daily release of the Brazilian official data.

The long-term projection is of great interest in a pandemic study. The general audience evaluating predictions is particularly interested in identifying some key aspects such as: the time in which the peak outbreak occurs, the time needed to reach a reasonably low daily count of cases/deaths and the errors associated with the estimated epidemiological curve. Developing a long-term projection, in a study where new observations are constantly included in each time series, is challenging and requires from the analyst some understanding about how to properly interpret the statistical results. The conclusions obtained from the current predictions may change significantly upon the arrival of new observations; see Chowell et al. (2020) for an additional discussion. The reader must note that the projection based on the data available until the current time point provides a static evaluation of the expected progress of the pandemic. This evaluation is done under the assumption that the epidemiological curve will not be influenced by any factor or event occurring in the future time points. Naturally, influential events cannot be avoided and the predictions will change throughout the study. Some examples of influential events are: increase/decrease in the number of tests to identify infected individuals in the population, level of social distancing in the population, governments adopting temporary restrictions for non-essential activities or the complete lockdown of cities. The presence of atypical observations, due to recounting procedures or delays to officially incorporate new cases in the data, can also impact the long-term projection. The take-home message here is that the statistical analysis should not be considered flawed given the dynamic context of the projections.

Predictions related to the daily number of deaths are more accurate than those associated with the number of cases. This aspect is justified by the fact that the under-reporting rate is certainly lower for the death count. In a pandemic where a large fraction of the population has mild symptoms, or it is even asymptomatic, many infected individuals will never be tested and they will remain as unknown cases during the outbreak. Obviously, this problem can only be alleviated by increasing the testing level to include not only the subjects having clear symptoms. In contrast, any death must be reported to the authorities and a detection test is expected to be part of the usual protocol required to issue the death certificate. However, some attention is necessary to avoid an excess death count associated with the target disease. As a simple example, consider an asymptomatic individual that dies in a car accident; although testing will indicate the infection, this death is not caused by the disease. Depending on the region where data is collected, it is possible that the local authority does not demand a detection test to be applied for every death. In other words, only suspicious cases, such as subjects that experienced breathing difficulties, will be formally tested to confirm the infection. If this policy is in effect for a given region, the under-reporting rate will be higher for the corresponding death counts. In summary, the statistical analysis in a pandemic study must be done with caution, i.e., the researcher must be aware that the reported observations contain some level of measurement inaccuracy. Please refer to the previous Chapters 2 and 7 for more details about data reporting problems. See also the references Veiga e Silva et al. (2020), Krantz and Rao (2020) and Lau et al. (2020) to see examples discussing issues in the context of COVID-19 data.

The CovidLP project was developed around a `Shiny` application providing an online web interface to show data and predictions corresponding to many regions in the world. `Shiny` (Chang et al., 2020) is a powerful tool within the software `R` (R Core Team, 2020) allowing an interactive statistical analysis for users. The analyst can easily select and explore the most recent results (short or long-term, death or new cases) for any available region. This dynamic investigation was extremely helpful for the task of daily monitoring the data and inferences. The `Shiny` visualisation tool greatly contributed to the fast identification of critical issues in the data leading to unreasonable predictions. The next sections illustrate some obstacles and challenges.

12.2   Atypical observations

As mentioned in the previous section, an important task of monitoring is to inspect the time series to identify atypical observations, especially those that may compromise the model fit or mislead the inferences of the epidemiological study (Chowell et al., 2020). The long-term projection receives great attention to evaluate the progress of the disease, therefore, one should always investigate the influence of an outlier on the main features of the estimated epidemiological curve. In the COVID-19 context, which is a large study involving many countries and regions, atypical count values may appear as a result of the many procedures chosen by the different governments to compile and report new cases or deaths.

One example of atypical observation is the detection of an inflated count with a magnitude higher than the one reported for all previous time points. This atypical case will be seen as a spike in the trajectory of the time series or a large jump in the monotone path of a cumulative curve. As an example, Figure 12.1 shows the occurrence of an inflated count near 22 March 2020 in Australia. Note that in general the curve is smooth and a jump representing a significant increase can be seen right after 21 March 2020. This behaviour can have many explanations, for example, it might be related to a data revision including previous cases that were not processed in the day of occurrence.

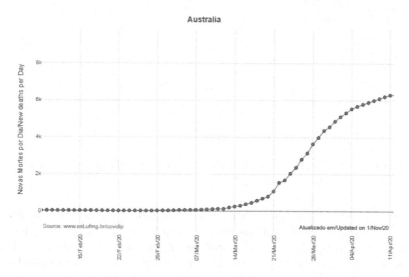

FIGURE 12.1: Cumulative curve of confirmed cases of COVID-19 in Australia per day, from February 2020 to April 2020.

Another example of an atypical count being observed for a region is a value that is too low, compared to the previous ones, representing a sharp decay in the trajectory of the time series of cases or deaths. During the COVID-19 pandemic, some regions exhibited a decay towards a negative value, which does not make sense in terms of possible response for a count variable. This situation may occur as a consequence of recounting procedures or changes in the way the data set is reported to the public. In a first inspection, the death of an individual may be confirmed as related to the target infection, but this status may change later if new evidence is found against that confirmation. Figure 12.2 shows the daily series of new COVID-19 cases in Spain. As can be seen, a negative value is reported on 24 April 2020, which clearly represents a drastic alteration in the trajectory of the sequence. As a result of this event, the curve displaying the number of cumulative cases will not be monotone in this case; a decay will occur between 23 and 24 April 2020. Again, this type of anomaly in the data may have serious implications for the model fit and for the conclusions drawn from the predictions.

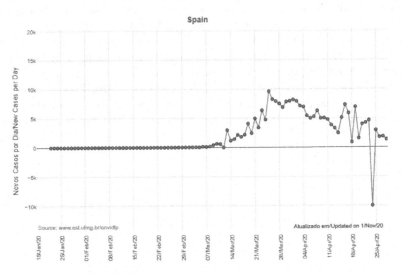

FIGURE 12.2: Number of confirmed cases of COVID-19 in Spain per day, from January 2020 to April 2020.

In summary, during an epidemic, the counts of cases may present fluctuations and atypical observations, such as negative counts. A possible form of exploring data in a more easy-to-analyse way without the erratic configuration, is to obtain the moving average of five or seven days, for example. This is a possible alternative for analysis, however, using the raw counts is appealing in terms of accounting for the observed variability. The next section discusses a specific scenario where an increasing-decreasing trend is found in different parts of the time series. This configuration is usually called a pattern of multiple waves and it is expected to occur in the sequential counts related to regions where the epidemic is only temporarily controlled. Many factors may explain an additional wave in the data pattern; mutations of the virus and a reduction in the level of social distancing are two of them. More details are discussed ahead.

12.3　Detecting multiple waves

There is no formal definition for an epidemic wave. The word "wave" suggests a rising number of cases or deaths, the presence of a peak and then a decay.

The term multiple waves brings the idea of a natural configuration of peaks and valleys in the time series. Many epidemiological models are built under the assumption of a single-peak outbreak, however, in practice, the course of several infectious diseases might unfold over time in somewhat predictable waves, with higher transmission rates in separated periods. The presence of this setting can be justified by different factors such as: authorities and people relaxing the level of social distancing after the first peak, the level of immunity in the population, mutations of the virus and the people's behaviour in each season of the year with a contrast between summer and winter. Further discussion about this topic can be found in Chowell et al. (2019) and Morens and Taubenberger (2009).

This section is devoted to the topic of monitoring the data set to detect multiple waves, taking the COVID-19 context as the major example. During the daily inspection of the sequential data explored in the CovidLP project, the first indication of a second wave occurred in late March 2020 for the count data corresponding to Iran. The evaluation task is done through visual inspection. Figure 12.3 shows that the series of confirmed cases in that country seems to reach a plateau, between 06–22 March 2020, followed by an abrupt growth. The model adopted to fit the data at the beginning of the project was a single-peak version being unable to handle this setting observed for Iran. Obviously, unreasonable inferences were obtained in this specific analysis and thus the best strategy was to temporarily remove, until model improvement, the Iran series from the list of countries explored via the online CovidLP Shiny application.

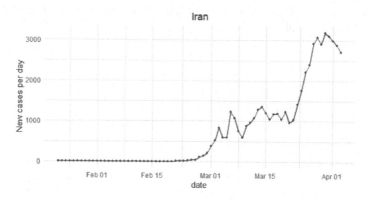

FIGURE 12.3: Number of confirmed cases of COVID-19 in Iran per day, from January 2020 to April 2020.

After Iran, more countries began to exhibit a behaviour that could be recognised as a second wave through the adopted criteria for visual inspection in the daily monitoring of the time series. As a consequence of this perception, it was necessary to develop and implement a model incorporating elements to deal with the presence of at least a second wave in different stages. Figure 12.4 shows the series related to Australia where one can clearly see two well-delimited and separated waves; the second wave reaches a higher peak compared to the first one. The pattern observed for Australia is not the same detected for other countries. Figure 12.5 shows the situation of Italy, where the first wave has a slow decay and the second wave, not yet finished in the example, indicates a fast increase that will lead to a higher peak of new cases. As a measure to control the pace of the second wave, the authorities in Italy decided to impose another severe period of restrictions to improve social distancing at the end of October 2020. The effect of the policy is not reflected in the trajectory of the series in Figure 12.5.

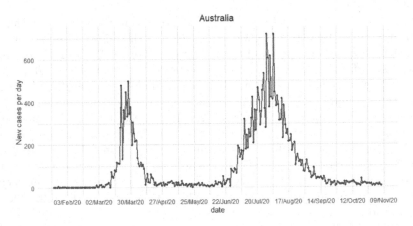

FIGURE 12.4: Confirmed cases of COVID-19 in Australia per day, from January 2020 to November 2020.

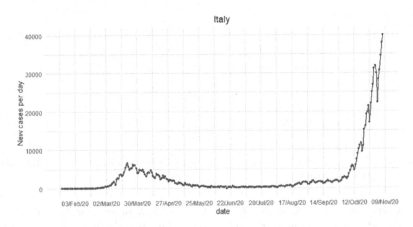

FIGURE 12.5: Confirmed cases of COVID-19 in Italy per day, from January 2020 to November 2020.

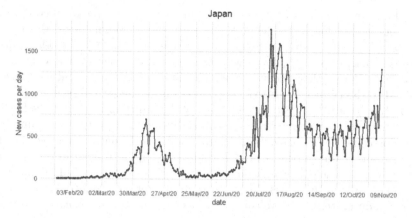

FIGURE 12.6: Number of confirmed cases of COVID-19 in Japan per day, from January 2020 to November 2020.

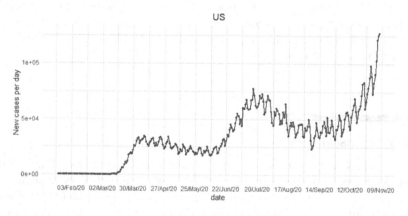

FIGURE 12.7: Number of confirmed cases of COVID-19 in the United States per day, from January 2020 to November 2020.

In addition to the second wave, after following the behaviour of the data related to various regions for many months, the visual pattern of a third wave was considered clear for some countries in the list investigated by the CovidLP project. Once again, an improvement was necessary to make the adopted model capable of accommodating the new feature. As one may expect, the model version designed for data series with three waves has a higher computational cost, which in turn imposes some difficulties for a daily update of the online CovidLP Shiny application. One possible solution to circumvent this issue is to assume a lag of a few days to update predictions of third-wave series. Figures 12.6 and 12.7 present the sequential data related to Japan and the United States; note that the third wave is in an initial stage for both cases.

The first wave of Japan is well delimited and the valley after the second wave does not reach a low level of counts near zero. The beginning of the third wave seems clear when looking at the increasing trend after 12 October. The United States scenario does not show a wave surrounded by valleys where counts are near zero. Here a slow decay is detected after the first peak and the growth related to the third wave is reaching values above the second peak.

In the CovidLP project, the detection of the time point in which the formation of a new epidemic wave begins is purely visual. Obviously, some numerical criterion could be established for this task, such as the one in Xu et al. (2020), for example. Even so, this strategy would still require the subjective specification of a threshold or condition. Note from Figures 12.4, 12.5, 12.6 and 12.7 that the shape of the waves differ substantially between countries, so there is no common pattern to establish a fixed decision criterion that can be applied to all possible scenarios.

Depending on the country, other features may be observed in the trajectory of the time series. The next section describes a specific aspect related to the rate of which observations are processed and incorporated into the official count. In other words, the weekday may interfere with the capacity to include a death/case in the database reported to the public. This is particularly true for Brazil and its states.

12.4 Seasonality

The previous section discussed how to detect multiple waves. This feature may be observed by the constant monitoring of the evolution of the counts of confirmed cases (or recovered, or deaths). The early recognition of this characteristic is important, so that governments can be prepared to take action. Also, it must be the case of adapting the model being used to fit the observations. The present section will illustrate the seasonality pattern that may appear in the data. The main aim here is to indicate a way of verifying this feature in the count series. Once this trend is detected and one wishes to include this information in the modelling procedure, Chapters 5 and 7 are recommended for further details. Here, the seasonality characteristic can be referred to as the delay in reporting, or as the "weekend effect". Chapter 5 discusses the causes of this peculiar aspect in the count data. The present section briefly explains it again and shows a simple way of detecting this pattern.

As mentioned above, the interest here is to analyse a possible systematic cyclical pattern in the data. In order to do so, consider Figure 12.8. The left panels show the number of new confirmed cases that occurred from 01 March 2020 to 07 November 2020 in Brazil (Fig. 12.8a), Costa Rica (Fig. 12.8c), and Paraguay (Fig. 12.8e). The panels on the right (12.8b, 12.8d, and 12.8f) are focused on the confirmed deaths for the mentioned countries in the same time period. In this figure, one can observe that the evolution of the counts has a very clear cyclical pattern composed of increasing counts followed by a substantial decay. The second half of the series for Brazil and Costa Rica indicate that after each strong drop, the higher level of the path is reached again, with the possibility of observing an even larger number of cases.

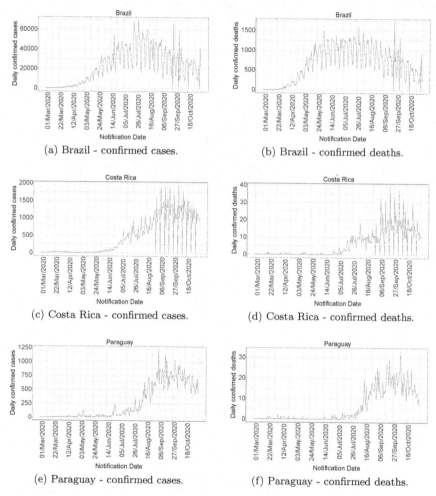

(a) Brazil - confirmed cases.

(b) Brazil - confirmed deaths.

(c) Costa Rica - confirmed cases.

(d) Costa Rica - confirmed deaths.

(e) Paraguay - confirmed cases.

(f) Paraguay - confirmed deaths.

FIGURE 12.8: Daily evolution of the confirmed cases (left panels) and confirmed deaths (right panels) of COVID-19 in Brazil, Costa Rica, and Paraguay. The data varied from 01 March 2020 to 07 November 2020 and they are ordered per reporting date. Data source: Brazilian Ministry of Health and database prepared by the Center for Systems Science and Engineering at Johns Hopkins University.

Further investigation led us to conclude that during weekends or holidays, there are fewer people working in all the procedures that involve testing and confirmation of cases. That is, fewer people to collect sample material, to analyse this material, as well as to insert data into electronic systems. Since there is, usually, less staff working in this entire procedure during weekends, there is also a lower count of reported cases on these days. One consequence of this situation is that, when considering cases by reporting date, one may observe a pattern that reflects the test timing, instead of the pure evolution of the disease. As a result, this pattern is not necessarily the same as the evolution by the date of first symptom, for example.

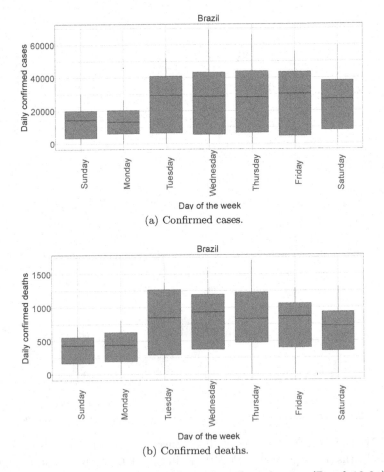

FIGURE 12.9: Boxplot of the number of confirmed cases (Panel 12.9a) and confirmed deaths (Panel 12.9b) of COVID-19 in Brazil, from 01 March 2020 to 07 November 2020, per day of the week. Data source: Brazilian Ministry of Health.

An alternative way of visualising this feature is to group data by the days of the week. This aspect is displayed in Figure 12.9. The graph presents the number of confirmed cases (Panel 12.9a) and confirmed deaths (Panel 12.9b), that occurred in Brazil, varying from 01 March 2020 to 07 November 2020 and grouped by each weekday. Similar to Figure 12.8, the reference date is the reporting date in this analysis. Each boxplot represents all observations related to a weekday during the entire period of study. Under this visualisation, it is no longer possible to interpret the evolution in time; instead, one can understand the distribution of the counts in each day of the week.

Figure 12.9a indicates that the median counts from Tuesday to Saturday are similar to each other. In addition, there is slightly less variability on Saturday. On the other hand, the entire distribution of the cases reported on Sundays and Mondays are similar among them, but persistently different compared to the previous mentioned days. Note that the interquartile ranges referring to Sundays and Mondays are lower than the medians of the other days. In addition, the variability of the distribution of cases reported in these two days is lower. The conclusion here is that there are lower reported cases on these days, and it is consistent throughout the time period under study. This pattern happens exactly due to less staff working on Saturday and Sunday, which impacts the reporting on the two subsequent days (Sunday and Monday, respectively). It is important to mention that this situation does not configure an under-reporting feature. These cases were reported, but with a delay regarding the date of first symptoms or testing result, for example. Figure 12.9b exhibits the distribution of the daily confirmed deaths with respect to the days of the week. The similarity is clear between this pattern and the one corresponding to the daily confirmed cases (Figure 12.9a). Note that the count values for the distribution related to Sunday and Monday are still low compared to the other days.

Figure 12.10 shows the boxplots per weekday for Costa Rica and Paraguay. Panels 12.10a and 12.10b are related to the confirmed cases and deaths in Costa Rica, respectively. Panels 12.10c and 12.10d are associated with Paraguay. In the analysis of Costa Rica, one can clearly note a discrepancy between the distributions of the cases reported on Sundays and the remaining days. The interquartile range is shorter for Sunday and the graph is located in a lower position. In contrast, the distribution of the counts are similar for all weekdays in Paraguay, see Panels 12.10c and 12.10d.

In the CovidLP project, the cyclical patterns of counts with decays occurring on the same weekday are referred to as seasonality, delay in the reporting, or "weekend effect". This pattern was also mentioned in Hsieh et al. (2010), Chowell et al. (2020), and Roosa et al. (2020). Chapter 7 and Marinović et al. (2015) show an interesting scheme illustrating possible steps and procedures that may cause a delay in the reporting process. Basically, this delay can be composed of a sum of other factors that may include the time between the infection and the first symptom, the need for medical prescription for an exam, testing procedures and their results, and, finally, the reporting part itself. It is important to mention that this effect is not exclusively restricted to Sunday and Monday. It will vary according to the regional culture of the place under investigation, among other factors.

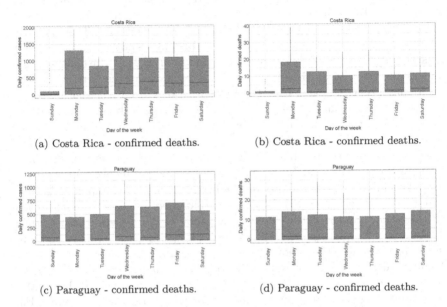

(a) Costa Rica - confirmed deaths.

(b) Costa Rica - confirmed deaths.

(c) Paraguay - confirmed deaths.

(d) Paraguay - confirmed deaths.

FIGURE 12.10: Boxplot of the number of confirmed cases and confirmed deaths of COVID-19 in Costa Rica (Panels 12.10a and 12.10b) and Paraguay (Panels 12.10c and 12.10d), from 01 March 2020 to 07 November 2020, per weekday. Data source: data-base prepared by the Center for Systems Science and Engineering at Johns Hopkins University.

In any real data analysis, one aims to build a model to interpret key aspects about the proposed study. As a result, elements are included in the modelling to allow capturing the necessary information. The aspects discussed in the present section are critical to improve predictions by avoiding the identification of an epidemiological curve that overestimates the number of cases/deaths. Ignoring the evidence of seasonality may lead to a misleading trajectory of decay in the evolution of cases. Thus, since this is a legitimate pattern in the data, including this feature in the modelling is of great interest for the research. Ideas on how to treat this characteristic were discussed in Chapters 5 and 7. The next chapter will resume this discussion by comparing the data against the results obtained with different ways of treating this characteristic.

Bibliography

Chang, W., Cheng, J., Allaire, J., Xie, Y. and McPherson, J. (2020) *shiny: Web Application Framework for R*. URLhttps://CRAN.R-project.org/package=shiny. R package version 1.5.0.

Chowell, G., Luo, R., Sun, K., Roosa, K., Tariq, A. and Viboud, C. (2020) Real-time forecasting of epidemic trajectories using computational dynamic ensembles. *Epidemics*, **30**, 100379.

Chowell, G., Tariq, A. and Hyman, J. M. (2019) A novel sub-epidemic modeling framework for short-term forecasting epidemic waves. *BMC Medicine*, **17**.

Dong, E., Du, H. and Gardner, L. (2020) An interactive web-based dashboard to track COVID-19 in real time. *The Lancet Infectious Diseases*, **20**, 533–534.

Hsieh, Y.-H., Fisman, D. N. and Wu, J. (2010) On epidemic modeling in real time: An application to the 2009 novel A (H1N1) influenza outbreak in Canada. *BMC Res Notes*, **3**.

Krantz, S. G. and Rao, A. S. S. (2020) Level of underreporting including underdiagnosis before the first peak of COVID-19 in various countries: Preliminary retrospective results based on wavelets and deterministic modeling. *Infection Control & Hospital Epidemiology*, **41**, 857–859.

Lau, H., Khosrawipour, T., Kocbach, P., Ichii, H., Bania, J. and Khosrawipour, V. (2020) Evaluating the massive underreporting and undertesting of COVID-19 cases in multiple global epicenters. *Pulmonology*.

Marinović, A. B., Swaan, C., van Steenbergen, J. and Kretzschmar, M. (2015) Quantifying reporting timeliness to improve outbreak control. *Emerging Infectious Diseases*, **21**, 209–216.

Morens, D. M. and Taubenberger, J. K. (2009) Understanding influenza backward. *JAMA*, **302**, 679–680. URLhttps://doi.org/10.1001/jama.2009.1127.

R Core Team (2020) *R: A Language and Environment for Statistical Computing*. R Foundation for Statistical Computing, Vienna, Austria. URLhttps://www.R-project.org/.

Roosa, K., Lee, Y., Luo, R., Kirpich, A., Rothenberg, R., Hyman, J., Yan, P. and Chowell, G. (2020) Real-time forecasts of the COVID-19 epidemic in China from February 5th to February 24th, 2020. *Infectious Disease Modelling*, **5**, 256–263.

Veiga e Silva, L., de Andrade Abi Harb, M. D. P., Teixeira Barbosa dos Santos, A. M., de Mattos Teixeira, C. A., Macedo Gomes, V. H., Silva Cardoso, E. H., S da Silva, M., Vijaykumar, N. L., Venâncio Carvalho, S., Ponce de Leon Ferreira de Carvalho, A. and Lisboa Frances, C. R. (2020) COVID-19 mortality underreporting in Brazil: Analysis of data from government internet portals. *J Med Internet Res*, **22**, e21413.

Xu, B., Cai, J., He, D., Chowell, G. and Xu, B. (2020) Mechanistic modelling of multiple waves in an influenza epidemic or pandemic. *Journal of Theoretical Biology*, **486**, 110070.

13

Investigating inference results

Vinícius D. Mayrink
Universidade Federal de Minas Gerais, Brazil

Juliana Freitas
Universidade Federal de Minas Gerais, Brazil

Ana Julia A. Câmara
Universidade Federal de Minas Gerais, Brazil

Gabriel O. Assunção
Universidade Federal de Minas Gerais, Brazil

Jonathan S. Matias
Universidade Federal de Minas Gerais, Brazil

CONTENTS

The reader is reminded that the monitoring context presented in Chapter 12 aims to identify and explain features in the observed time series that will clearly impact the analyses based on predictions. In contrast, the present chapter is focused on describing some inconsistencies detected when evaluating inference results provided by the version of the Bayesian model applied to the daily updated COVID-19 data available in each stage of the study. Some scenarios with unreasonable predictions are discussed to illustrate problems appearing in different parts of the pandemic period. Naturally, the identification of an inference issue imposes the need for improving the Bayesian model, or the computational strategy, in order to handle the obstacle. The task of

DOI: 10.1201/9781003148883-13

regularly monitoring predictions contributed to a study where the adopted model evolved to better accommodate the sequential data and to avoid unrealistic interpretations. The examples discussed here may also be observed in other pandemic (or epidemic) studies.

13.1   Monitoring inference issues

Predictions related to the epidemiological curve are a central aspect in a pandemic (or epidemic) study. The analysis of this element provides key information to evaluate the progress of the target disease in the population, such as identifying the expected time point corresponding to the outbreak peak, establishing the duration of the pandemic by determining the period in which the number of cases/deaths will become too low, estimating the number of infections or deaths on any given time point and, finally, determining the total number of infections or deaths by the end of the outbreak. Obviously, in a Bayesian study, the estimated epidemiological curve is summarised from a posterior distribution. The variability associated with this distribution must be considered in the analysis, through prediction intervals, to express the uncertainty related to the trajectory of the mean curve. The reader must also note that the prediction on a given time point represents the expected behaviour of the curve conditioned on the adopted statistical model—and its assumptions—and the data available in that moment. The predictions are established by also assuming that the pattern of the time series will not show a drastic change due to influential events or atypical observations occurring in the upcoming time points.

Two scenarios can be considered for predictions: the short and the long-term. The accuracy is higher when making inferences about the values to be observed a few time points ahead (short-term), which is a known result in time series analysis. Although this feature makes the short-term predictions attractive for an accurate statistical evaluation, the long-term predictions, having higher uncertainty for distant time points, are in general the focus of those interested in studying the progress of a disease. The reason for this is the fact that long-term predictions allow a more complete investigation by identifying important aspects such as the position of the peak and the time required to a decay. As a result of the higher uncertainty associated with the long-term scenario, unreasonable configurations may occur based on the estimated epidemiological curve. In this case, monitoring has an important role in the development of a project where long-term predictions are daily updated due to the arrival of new observations.

In many situations, due to the complexity of the adopted model, the Bayesian approach can only be used through the implementation of a sampling strategy to indirectly generate values from the joint posterior distribution.

This target posterior is only known up to a normalising constant, which is an obstacle for direct sampling. Markov Chain Monte Carlo (MCMC) methods are widely used to promote the mentioned indirect sampling. This class of methods comprises different computational algorithms such as the Gibbs Sampler (Geman and Geman, 1984; Gelfand and Smith, 1990), the Metropolis-Hastings (Hastings, 1970; Metropolis et al., 1953) and the NUTS (No-U-Turn) sampling based on Hamiltonian dynamics (Hoffman and Gelman, 2014). See further details about MCMC in Section 6.2.2 of Chapter 6. These algorithms can be used separately or in a joint setting, for example, the Gibbs Sampling with Metropolis-Hastings steps. The choice of the algorithm depends on the chosen statistical model and also on the computational time required to handle the associated simulations. Great computational efficiency in ordinary computers is appealing for the general public. Any computational method may suffer, in some applications, with numerical instability occurring due to problematic values with a too small or too large magnitude appearing throughout the internal calculations of the algorithm. This issue can only be circumvented by implementing smart computational strategies to avoid the small/large numbers. Generally, the problem will be detected through an error/warning message reported by the software running the MCMC. The problem can also be manifested through the lack of convergence of one or more chains obtained from the MCMC. Consequently, poor convergence will lead to a strange or inconsistent result when the posterior estimates are summarised. In view of these aspects, note that it is essential to monitor the estimates (and predictions) keeping in mind that a possible cause of the unreasonable result is the lack of convergence of the MCMC algorithm.

13.2 Monitoring and learning

The constant monitoring of predictions is important to understand the development of the epidemic and to evaluate the performance of the current modelling approach. In the specific context of the CovidLP project, through the daily incorporation of new observations, it was possible to see from visual inspection that, for many regions, the decreasing part of the epidemiological curve seemed to be longer than the increasing period. In other words, at the beginning of the study, most countries were displaying a one wave configuration with a fast curve increase before the peak and a slow curve decrease after the peak. According to this perception, the strategy chosen in the analysis was to impose an asymmetric shape on the epidemiological curve representing the average number of new cases or deaths $\mu(t)$. The basic idea adopted in this situation is to assume the condition $M(t_{peak}) \leq M(t_{end}) - M(t_{peak})$, where $M(t_{peak})$ and $M(t_{end})$ are the cumulative number of cases/deaths until the peak and the final time point, respectively. Further details about imposing an asymmetric shape on the mean curve can be found in Chapter 3. Chapter 6 discusses this aspect under the Bayesian point of view.

Another important aspect identified during the CovidLP project was the fact that unreasonably large quantities were estimated based on the long-term predictions for some countries, and also for Brazilian states. This includes the predicted number of cases/deaths to be reported in a single day and also the predicted cumulative number of cases/deaths to be observed at the end of the pandemic. The size of prediction intervals, where an unrealistic upper limit is observed, is another issue to be highlighted. In order to judge the magnitude of the predictions as inappropriate, one must consider the population size of the region under study. For instance, the cumulative number of cases/deaths are not expected to be larger than or near the total number of individuals living in the region. In this scenario, it was clearly necessary to verify the data for atypical values, the convergence of the MCMC, the adequacy of the proposed model, the choice of priors, and other elements directly impacting the predictions. In terms of modelling improvement, a modification to handle the magnitude problem was to impose a truncation establishing an upper limit for the predicted results. The chosen upper limit, for the count time series of new cases, was 10% of the population size. Other values may be suitable, e.g., 20% or 50%. This value can be lower for the time series of deaths, since only a fraction of the infected individuals will die. Larger percentages (say $> 50\%$) do not make sense in epidemiological studies. In order to define the mentioned percentage for the time series of deaths, one can simply evaluate the daily proportions of the number of deaths divided by the number of cases. Consequently, if assuming that 10% of the population will be infected and that 2% of the infected cases will die, consider $0.02 \times 0.10 = 0.002 = 0.2\%$ as a possible choice for the truncation limit. Other percentages can be explored for the series of death cases.

Again, the upper bounds of the prediction intervals should also make sense in the analysis. In contrast with the posterior mean and median, a truncation is not imposed to determine the intervals. In order to avoid displaying a prediction interval whose upper limit is not consistent with the population size, the straightforward strategy is to omit the interval from the predictions released to the general public. The researcher must monitor the updated predictions, to be obtained in the upcoming days, to check if the problematic upper bound changes. Figure 13.1 illustrates, in the COVID-19 context, the epidemiological curve (posterior mean) obtained for daily confirmed cases in Bolivia. Observed counts are available until 24 May 2020, and the estimated peak reaches a number of cases 3.5 times higher than the largest observed count in the series. The posterior uncertainties are high for the predictions in this case, which leads to prediction intervals with large amplitudes. This graph was published in the CovidLP shiny application without the problematic 95% intervals that are difficult to interpret from the epidemiological perspective. The reader must note that the CovidLP project applies the same model to fit many time series with different characteristics. The mentioned upper limit issue only occurs in very few cases and mostly when few counts were available. This problem tends to disappear upon the arrival of a new observation on the next day.

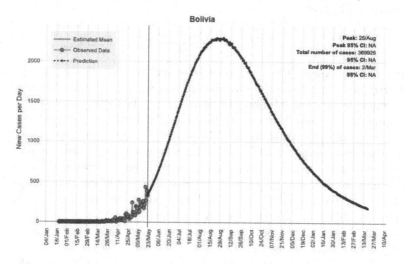

FIGURE 13.1: Long-term predictions for the number of confirmed cases (COVID-19) in Bolivia using data observed until 24 May 2020.

13.3 Evaluation metrics

The visual inspection to identify features suggesting unrealistic predictions, as discussed in the previous section, is not the only way to monitor results on a daily basis to constantly report the progress of an epidemic. This task can also be developed by exploring some formulations or metrics that use the predictions to provide quantitative responses to support the analysis intended to evaluate and compare models. The main aim of the present section is to indicate some options considered during the CovidLP project; some metrics mentioned here were not applied to the project.

Given that the short- and long-term predictions are the main outcomes when studying an epidemic, a natural choice to define a model performance measurement is to use the difference between a prediction and the corresponding real observation. In other words, after reporting the predictions using the data until day t, the researcher will calculate the distance between each prediction and the corresponding observation for the days $t+1, t+2, t+3$ and so forth. Note that the time point related to the last observed count should be well defined for this analysis. Greater distance with respect to the real value is expected for the prediction related to $t + k$, where k is a large positive integer. In line with the previous idea, the first metric to be defined is the Relative Error (RE) given by:

$$\text{RE}_i = \frac{\text{pred}_i - \text{obs}_i}{\text{obs}_i}, \tag{13.1}$$

where obs_i represents the true count observed for the i-th day and pred_i is the predicted value (e.g., the posterior mean) for the i-th day. Alternatively, one may calculate the Absolute Relative Error (ARE) by simply computing the absolute difference between the true count and the predicted value, that is $\text{ARE}_i = |\text{pred}_i - \text{obs}_i|/\text{obs}_i$. Both RE and ARE can be interpreted as percentages if results are multiplied by 100. The presence of obs_i in the denominator indicates that the magnitude of the distance in the numerator is reported in terms of how big it is with respect to the magnitude of the true count. This configuration has been adopted to evaluate bias in the literature; see, for example, Almeida et al. (2018), Mayrink et al. (2020), and Almeida et al. (2021). Note that if $\text{RE}_i > 0$, the predicted value is greater than the true count, meaning overestimation. In contrast, $\text{RE}_i < 0$ means an underestimation. Note also that $\text{ARE}_i > 0$, since the absolute value is considered for the numerator and the denominator is a positive count. In the presence of zero counts, one may set $\text{obs}_i + 1$ and $\text{pred}_i + 1$ in the formula to avoid computational issues.

The main difference between ARE and RE is related to whether the signal of the difference in the numerator is relevant or not for the analysis. In a comparison study, it can be more convenient to explore the ARE, since the key information is the magnitude of the error and not the indication of over- or underestimation. As described in Chapter 12, atypical observations may appear in the time series and it will impact the proposed metric (13.1); $100 \times \text{ARE}_i \approx 100\%$ would be obtained. In the analyses of the CovidLP project, ARE and RE are adopted as the main metrics.

Alternatively, one may choose the Mean Squared Error (MSE) to summarise the predictive performance of the epidemiological model. In a study where predictions are determined based on a time series with observations until day t, the MSE is as follows

$$\text{MSE}(t, T) = \frac{\sum_{i=1}^{T}(\text{pred}_{t+i} - \text{obs}_{t+i})^2}{T}. \tag{13.2}$$

In this case, the result is a single positive number summarising the squared magnitude of distances with respect to the true count for a period involving T days after day t. Obviously, the observations after day t must be available to allow this calculation; again, these elements are not used to fit the model. The presence of an atypical value will impact the MSE, leading to a large value due to the presence of obs_{t+i} only in the numerator of (13.2). For additional details see Roosa et al. (2020).

The next measurement to be discussed in the present section has a formulation quite similar to the MSE in (13.2). The Mean Absolute Error (MAE) is also built using predictions from a model fit related to a time series with observations until day t. The outcome is a single positive number representing the average absolute distance between predictions and observations related to the period of T days after t. This choice of error measurement is given by

$$\text{MAE}(t, T) = \frac{\sum_{i=1}^{T} |\text{pred}_{t+i} - \text{obs}_{t+i}|}{T}. \tag{13.3}$$

It is important to highlight that the observations obs_{t+i} are only used to calculate the MAE. In other words, the data is separated in a training and a test set. The training set is the time series containing the count data to fit the model. The test set is composed by the observations being compared to the predictions. The result from (13.3) is less impacted than the MSE by the presence of an outlier in the time series. Note that the square in the numerator of (13.2) will impose a large number within the sum for the outlier case; the corresponding value in the MAE will have lower magnitude. In summary, the MAE would be a better choice, with respect to the MSE, for the analysis related to a time series where atypical values are detected.

The last comparison measure to be mentioned in this section is the coverage percentage (CP). This metric is useful for evaluating prediction intervals (PI). Under the Bayesian framework, a PI of 100 $p\%$ indicates a probability p that the true observation is in the interval of the prediction. Usual values for p are 0.90, 0.95 and 0.99. Therefore, the CP is computed by obtaining the percentage of intervals that include the corresponding real value. Ideally, the CP is a percentage close to the nominal level 100 $p\%$.

The authors in Kolassa (2020) indicate that the choice of an appropriate error measurement depends on the type of point estimate (e.g., mean or median) being used. One may also prefer to have the evaluation based on the full predictive distribution, rather than focusing on point estimates; see Funk et al. (2019). These references provide important insights about assessing the performance of epidemic forecasts, however, they do not indicate that the metrics presented here are useless. On the contrary, these metrics are useful in the context of the book. In particular, exploring the full predictive distribution for several time points, in a study involving many time series with a daily resolution, is computationally difficult to handle.

13.4   Practical situations

The previous section focused on details about comparison metrics that can be used to evaluate the quality of predictions across the time. The constant evaluation of these outcomes is important due to several reasons, such as the adequacy and the performance of the model being applied. Moreover, it may provide indications of features to be included in the modelling approach. Given the importance of the mentioned metrics, the present section is dedicated to illustrate their usage. In addition, graphs will be suggested to examine the results.

Predicting the number of new cases on the fly during an epidemic is very challenging. Just to cite a few reasons, the definition of a confirmed case may change throughout the epidemic, and data can have fluctuations that may seem an indication of a situation after the peak. In addition, the values can represent an aggregation of cases coming from heterogeneous regions, which can lead to larger data variability. Also, the modelling approach may have modifications and upgrades across the time. The authors in Chowell et al. (2020) reinforce the difficulties in this task.

Taking into account the topics discussed above, the next subsections suggest how to use comparison metrics to evaluate predictions. Most of the comparisons in the discussion are based on the daily updated outcomes from the CovidLP project and the real data available for the corresponding days. They were separated into three main themes: for an overall comparison, see Subsection 13.4.1; to investigate a possible seasonality pattern comparing data and results of the modelling approach, go to Subsection 13.4.2; finally, if the aim is to observe a possible disconnection between a modelling approach based on a growth curve and a multiple wave scenario, read Subsection 13.4.3. Nonetheless, it is important to mention that the next discussions are only suggestions to develop the analysis. The reader may explore other variations for a data-specific scenario.

13.4.1   Overall comparison

This subsection is focused on exploring the comparison metrics, described in Section 13.3, in an overall evaluation of predictions against the real count values. The projection for the number of confirmed cases of a disease in an epidemic may have variations as new observations arrive throughout the time. These variations are a natural consequence of changes in social distancing measures and other actions to control the epidemic. As an example, consider the number of confirmed cases of COVID-19 that occurred in Argentina, from 23 January 2020 to 31 October 2020. The time series can be seen as grey dots connected with lines in Figure 13.2. This graph illustrates the variation of the predicted mean curve $\mu(t)$ across the time. Each black line represents the predicted mean curve considering data until a specific date; the final time points are: (i) 14 June, (ii) 10 July, (iii) 02 August, (iv) 29 August, (v) 05 October, and (vi) 28 October 2020. These options were chosen to have a spaced setting over the period of time under investigation; no particular reason is considered to select any of them.

The first aspect to have in mind in this analysis is the fact that the predictions reflect the most probable epidemiological curve, which is obtained with the available data and assuming fixed conditions related to lockdown, social distancing, testing rates, etc. If any of these conditions change, the estimated curve can show alterations. Now, turning attention back to Figure 13.2, note that when accounting for the time series until 14 June (black solid line), a rapid and consistent increase is detected for the predicted number of cases. However, the true growth of the data (grey dots) turned out showing a slower increase than the solid curve. Yet, predictions considering data until 10 July, 02 August and 29 August tended to underestimate the trajectory of the epidemiological curve, as the number of cases (grey dots) are found above the predicted curves. Finally, consider the curve based on the counts until 28 October. If the mentioned conditions are kept the same for the remaining time points, the pandemic is expected to have reached the peak in October. As a final comment about this illustration, note that the main goal here is not to choose or recommend one of the predicted curves. The central aspect of this analysis is to display the different paths obtained for the prediction curve as new observations are incorporated in the time series.

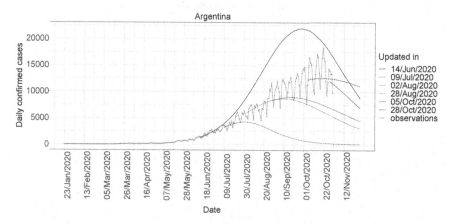

FIGURE 13.2: Comparison of the prediction evolution for the number of new confirmed cases (COVID-19) in Argentina considering six periods. The initial date is 23 January 2020 for specifications. The final dates are: 14 June, 10 July, 02 August, 29 August, 05 October, and 28 October 2020. Data source: Center for Systems Science and Engineering at Johns Hopkins University.

In what follows, Figure 13.3 shows the relative error for both short-term (Panel 13.3a) and long-term (Panel 13.3b) predictions of the number of COVID-19 confirmed cases in Argentina. Here, in both panels, the horizontal axis represents the date on which the projections were updated. The short-term prediction concerns the outcomes until seven days ahead. Then, each boxplot in Panel 13.3a shows the relative error using predictions and true counts associated with the mentioned period. The solid black line represents the reference value zero, meaning no difference between estimates and true values. A general conclusion is that the reference 0 can be found within many boxplots, especially the last ones on the right. This suggests an existing

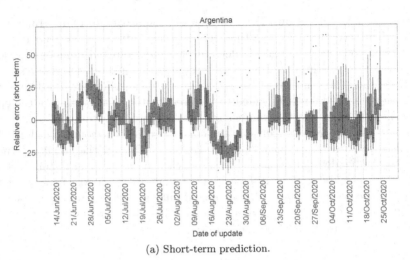

(a) Short-term prediction.

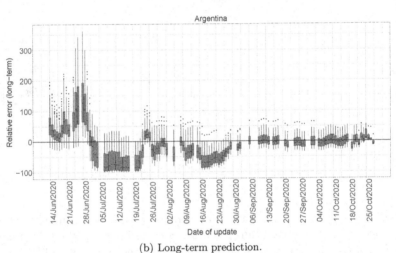

(b) Long-term prediction.

FIGURE 13.3: Relative error for both short- (a) and long-term (b) predictions related to the number of new confirmed cases (COVID-19) in Argentina. The analysis is based on data from 23 January 2020 to 31 October 2020. In this version, the absolute value is not considered for the numerator of the RE. Data source: Center for Systems Science and Engineering at Johns Hopkins University.

variation around 0 with the median being near the reference. In addition, note that the scale ranges from −40 to 60, therefore, the model tends to slightly overestimate rather than underestimate the confirmed cases. A comparison in terms of over/underestimation can also be done through Figure 13.2.

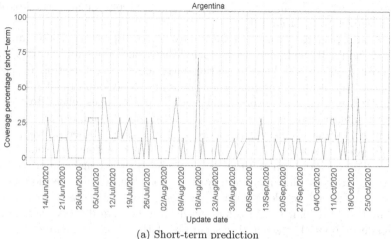

(a) Short-term prediction

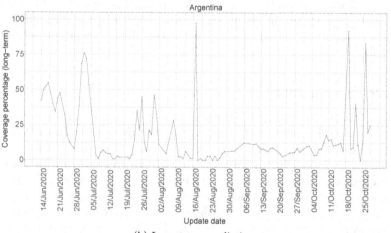

(b) Long-term prediction.

FIGURE 13.4: Coverage percentage for both short- (a) and long-term (b) predictions related to the number of new confirmed cases (COVID-19) in Argentina. The analysis is based on data from 23 January 2020 to 31 October 2020. Panel 13.4a concerns the predictions for only seven days ahead. Panel 13.4b accounts for all days ahead until the end of the pandemic. Data source: Center for Systems Science and Engineering at Johns Hopkins University.

When it comes to the long-term predictions, see Figure 13.3b. This graph shows the relative error for all upcoming days until 31 October 2020. From left to right, each boxplot is built with one less observation. It is worth pointing out here that the scale of the vertical axis is larger than that in Panel 13.3a. In addition, as time passes, the time series increases (i.e., more information is available) and the number of days to be predicted reduces (approaching 31

October, which is the final prediction date). As a consequence, the boxplots tend to concentrate around 0.

A final aspect to consider in this analysis is the prediction interval. This interval should be wide enough to include the true count and short enough to express accuracy and lower posterior uncertainty. An interesting discussion about the amplitude of the prediction interval can be found in Chapter 4. As an illustration, Figure 13.4 shows the coverage percentages related to the 95% Bayesian interval built for the predictions regarding the number of COVID-19 cases in Argentina. The coverage percentage is obtained based on the proportion of days in which the interval captures the corresponding true count. Panel 13.4a refers to short-term and Panel 13.4b concerns the long-term projections. One can clearly see that many intervals are too short to include the true values for both short- and long-term predictions. Note that low percentages are exhibited in both panels and some few days indicate a 0% coverage. Again, the reader is invited to see a discussion about this topic in Chapter 4.

This section was devoted to show how to use comparison metrics to evaluate predictions in an overall analysis. Monitoring this type of result is important as it gives an insight about adaptations needed in the Bayesian model. The next subsection investigates a possible seasonality pattern in the sequential data.

13.4.2   Seasonality

In the COVID-19 context, the seasonality is given by a cyclical pattern comprising consecutive periods of seven days in the time series. In the CovidLP project, this behaviour is called the "weekend effect" occurring due to delays in reporting. A modelling approach that includes this feature was discussed in Chapter 5. In addition, Chapter 12 indicates how to identify the pattern based solely on the visual inspection of the observed data series. The present section puts together the concepts described in the two mentioned chapters and compares estimates with true values. The main aim is to illustrate how comparison metrics can be useful to support the decision of including a seasonality effect or to consider alternatives such as changing the data to a weekly scale. This decision is case-specific and it tends to improve the results.

In order to develop the mentioned comparison, the data series of two Brazilian states are studied: Minas Gerais (MG) and São Paulo (SP). The data are the daily number of COVID-19 new confirmed cases registered between 25 February 2020 and 31 October 2020. The analysis is conducted by fitting: (*i*) the regular model assuming the mean growth curve as the generalised logistic, (*ii*) a similar alternative including the "weekend effect" related to Sundays and Mondays, and (*iii*) the first modelling specification but grouping data to a weekly scale. The original time series and the mean curves obtained via approaches (*i*) and (*ii*) can be seen in Figure 13.5.

In Figure 13.5, the grey dots connected with lines represent the observed counts. The dashed black line is the mean curve from the regular model and

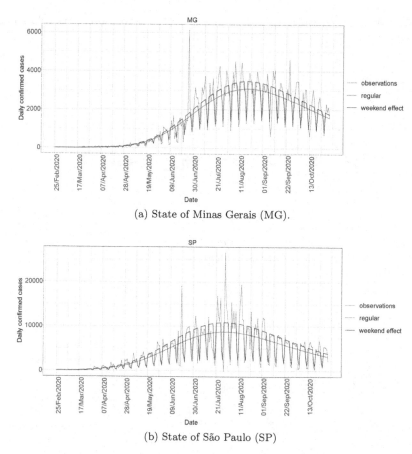

(a) State of Minas Gerais (MG).

(b) State of São Paulo (SP)

FIGURE 13.5: Comparison of mean curves $\mu(t)$ based on the results from the regular (dashed black line) and the weekend effect (solid black line) models. Time series of COVID-19 confirmed cases in the Brazilian states of Minas Gerais (a) and São Paulo (b). Data ranging from 25 February 2020 to 31 October 2020. Data source: Brazilian Ministry of Health.

the black solid line represents the mean curve including the seasonal effect. Hereafter, the regular and seasonal model fits will be referred to as "Model fit 1" and "Model fit 2", respectively. The main goal in this illustration is not to provide a clear distinction between the two mean curves. In fact, the study is intended to evaluate the impact of adding parameters to accommodate seasonality. In this perspective, note that the mean curve obtained via Model fit 1 is lower than the main trajectory related to Model fit 2. This result is intuitive since the model in Model fit 2 is more flexible to establish a trajectory with a fast decay/growth behaviour to handle the lowest observations. Once these values were accommodated by the seasonal effect, the mean curve is able to show a general path that is better adapted to the counts related to Tuesday–Saturday.

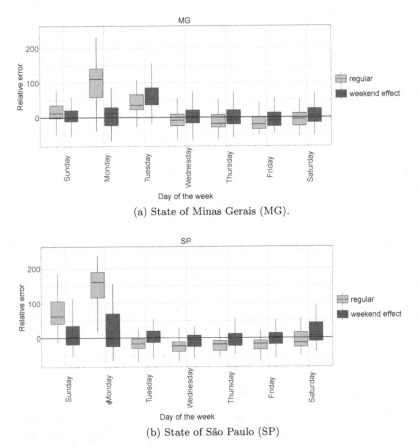

(a) State of Minas Gerais (MG).

(b) State of São Paulo (SP)

FIGURE 13.6: Comparison of the relative errors based on the results from the regular and the weekend effect models. Predictions for COVID-19 confirmed cases in the states of Minas Gerais (a) and São Paulo (b), Brazil. Outliers were removed from the boxplots to reduce scale and improve visual inspection. Data ranging from 25 February 2020 to 31 October 2020. Data source: Brazilian Ministry of Health.

Figure 13.6 compares Models 1 and 2 through the relative error for both Minas Gerais (Panel 13.6a) and São Paulo (Panel 13.6b). In this case, the RE is grouped by the day of the week in which the COVID-19 cases were registered. In other words, each boxplot represents the RE metrics corresponding to a specific day of the week. The boxes in light and dark grey represent Model fit 1 and 2, respectively. In addition, the horizontal solid black line indicates the reference value 0, which represents the level where estimate and true value are the same. Note that the sequential evolution of the RE throughout the time cannot be studied using these graphs. The analysis here is only intended to explore model performance for each day of the week.

Figure 13.6a shows that Model fit 2 provides a median RE closer to the reference value 0 than Model fit 1, for all days of the week, except Tuesday and Saturday. It is also noteworthy that even though the seasonal effect on Sunday provided a better fit, it seems that Tuesday has a stronger effect on the reporting delays. Therefore, the results here seem to indicate that it could be an interesting strategy to include another seasonal effect on Tuesday. Alternatively, one could consider the weekend effect restricted to Monday and Tuesday. Nevertheless, the model in Model fit 2 seems to have better performance than the one in Model fit 1.

Figure 13.6b clearly indicates better performance of the model in Model fit 2 with respect to the one in Model fit 1. The median RE from Model fit 2 is closer to 0 for all days of the week. Note that high variability is exhibited by the wide boxplots related to Sunday and Monday. This is a strong indication that setting the seasonal effect for Sunday and Monday is an appropriate choice for the São Paulo data.

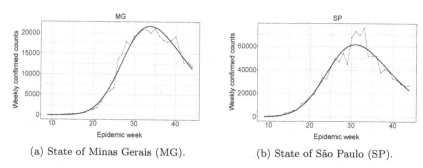

(a) State of Minas Gerais (MG).　　　　(b) State of São Paulo (SP).

FIGURE 13.7: Weekly new confirmed cases of COVID-19 along with estimated mean curve. Panel 13.7a concerns the state of Minas Gerais. Panel 13.7b is related to the state of São Paulo. Data ranging from epidemic week 9 to 44. Data source: Brazilian Ministry of Health.

Another strategy suggested here is to consider data as the moving average or to group the series of counts in weekly or even epidemic-weekly scale. This strategy may be challenging, or even unfeasible, to be put into practice when there are few counts or when the details of a daily resolution is more interesting, in terms of information, for the target audience. For example, two months of data as daily counts lead to approximately 60 observations; but only 8 in a weekly scale. It can also be the case that there are more parameters to estimate than data points available, which brings another difficulty to model the epidemic data. Notwithstanding, a drawback of this procedure is the loss of precision since projections would now be on the same grouped scale. Then, one would not know what to expect in one or two days ahead, but one or two weeks ahead instead. In the literature, there are works making use of both strategies: Hsieh and Cheng (2006), Hsieh et al. (2010), Chowell et al. (2019), Roosa et al. (2020), Schumacher et al. (2021) modelled data in a daily scale, whereas, Hsieh and Ma (2009), Hsieh and Chen (2009), Hsieh et al. (2010), Wang et al. (2012), Pell et al. (2018), Chowell et al. (2019), Funk et al. (2019) chose the weekly (or epidemic weekly) resolution.

As an illustration, the data explored in Figure 13.5 were grouped by epidemic week and then the regular model was fitted to this modified series. The regular model is the one for which results are shown as dashed black curve in Figure 13.5 and as boxplots in light grey in Figure 13.6. The results related to the weekly scale fitting will be called "Model fit 3" and they are seen in Figure 13.7. Panel 13.7a shows the weekly counts, and the respective fitted mean curve, for the new confirmed cases of COVID-19 in the state of Minas Gerais. Panel 13.7b is related to the state of São Paulo. The visual inspection indicates that the mean curve seems to accommodate the data very well. In addition, it is worth mentioning that the weekend effect is no longer present due to the sum of the counts within each epidemic week.

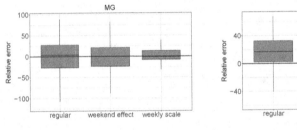
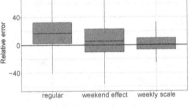

(a) State of Minas Gerais (MG).   (b) State of São Paulo (SP).

FIGURE 13.8: Comparison based on the RE calculated from fitting the regular model to daily resolution data, the model with weekend effect to daily resolution data, and the regular model to weekly resolution data. Outliers were removed from the boxplots to reduce scale and improve visual inspection. Panel (a) represents Minas Gerais and (b) is related to São Paulo. Data ranging from 25 February 2020 to 31 October 2020 (equivalent to epidemic week 9 to 44). Data source: Brazilian Ministry of Health.

Figure 13.8 shows the RE for the three mentioned model fits applied to the time series of Minas Gerais (Panel 13.8a) and São Paulo (Panel 13.8b). Note that this comparison has some limitations because, even though the calculation of the RE takes into account the observations, data in the third boxplot are not in the same scale as in the first and second. Consequently, the number of observations to build the third boxplot is not the same as the number related to the first and second graphs. Also, the model in Model fit 2 (weekend effect) is different from the others (regular modelling). Taking into account all these aspects, in the analysis of Panel 13.8a it is clear that the median RE from Model fit 1 is slightly above the reference value of 0. Moreover, values of RE are less dispersed in Model fit 2, and even less dispersed in Model fit 3. The same conclusions can be obtained from Panel 13.8b, although deviation from 0 is more accentuated for both Model fits 1 and 2.

The discussion in this section can be summarised with the conclusion that the RE is a useful evaluation tool when seasonality is present in data. In the analysis of the Minas Gerais data, the results from Model fit 2 suggested an extra day with seasonal effect, which indicates that the model should be modified to improve performance, or the data should be grouped to apply a regular approach. On the other hand, the analysis of the São Paulo data confirmed the chosen configuration of seasonal effect. Moreover, it is important to highlight that this is just an illustration about using the RE metric. Other alternatives can be tested by changing the mean curve function (see Chapter 3) or changing the distribution attributed to data (details in Chapter 4). All metrics discussed in Section 13.3 can be useful to explore the results. In terms of data scale, if one seeks to identify the days of the week in which lower counts are often reported or if the daily scale is preferable, then the weekend effect model is a good option. However, grouping data and changing the scale is an alternative and the performance of the approaches may be data-specific.

13.4.3 Multiple waves

The detection of multiple waves based on visual inspection of the data was a topic previously discussed in Chapter 12. The goal of the present section is to indicate some challenges related to predicting cases/deaths in a multiple wave scenario. Initially, a model assuming a single peak will be explored to evaluate discrepancies between predictions and true observations, i.e., predictions indicate the end of the epidemic in a situation where another wave is beginning with the count values increasing again. The pace of the epidemic progress differs from region to region, therefore, periodical monitoring of predictions is absolutely important to identify any alteration in the expected trajectory of the time series.

Figure 13.9 shows the predictions (solid curve) related to the number of confirmed cases (COVID-19) in Russia. The adopted model assumes a single wave, but the time series (grey dots) clearly indicate two increasing periods. As a result, the epidemiological curve does not represent the path of the data and the growth observed between September and October is not captured by the model. The identification of an additional wave is purely visual in the analyses of the CovidLP project. The daily monitoring is important for this evaluation. The detection of MCMC convergence issues and the absence of prediction intervals around the mean curve are perhaps the first signs of inappropriate model fitting due to multiple waves.

Figure 13.10 shows another example in a context related to the number of confirmed cases (COVID-19) in Japan. In early October 2020, the Japanese

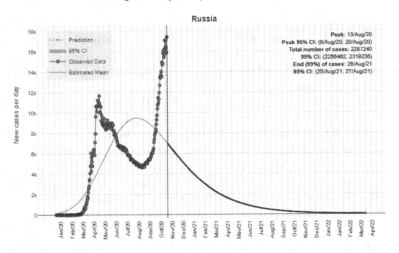

FIGURE 13.9: Predictions for number of confirmed cases (COVID-19) in Russia. Model assuming a single wave in this case. Data source: Center for Systems Science and Engineering at Johns Hopkins University.

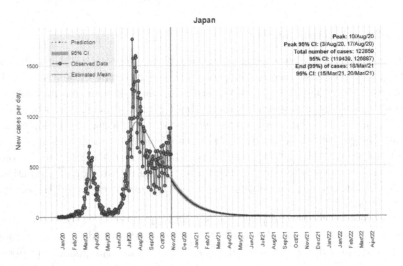

FIGURE 13.10: Predictions for the number of confirmed cases (COVID-19) in Japan. Model assuming two waves in this case. Data source: Center for Systems Science and Engineering at Johns Hopkins University.

time series seems to indicate the beginning of the third wave. The model adopted in this case was modified to account for two waves, however, this choice is inappropriate to handle more than two peaks. Note that the solid curve is well adapted to the first two waves, but the predictions have a decreasing path as opposed to the increasing trend observed in the last part of the series. This analysis provides a clear indication that the model should change to deal with a third peak. Apart from the contrasting behaviour of the predictions and the true counts, the narrow prediction interval (grey shade) around the mean curve can be seen as an indication of a problematic fitting.

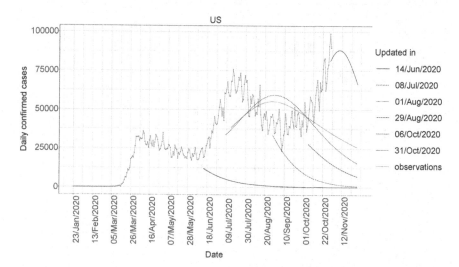

FIGURE 13.11: Comparison of the mean curve $\mu(t)$ representing predictions for the number of confirmed cases (COVID-19) in the United States. The initial date of the times series is 23 January 2020. The last dates of the series are: 14 June, 09 July, 01 August, 01 September, 06 October, and 31 October 2020. Data source: Center for Systems Science and Engineering at Johns Hopkins University.

The previous illustrations reinforce the importance of constantly monitoring the model outcomes. The evaluation of the proximity between predictions and true counts may give an insight about the beginning of an extra wave of infections or deaths. In particular, the progress of the COVID-19 time series in the United States configured a challenging scenario to be modelled; see Figure 13.11. In the CovidLP project, four different models were applied throughout the study until 31 October 2020. First, a single wave growth curve model was used, since most regions under investigation were in the initial stage of the pandemic with only one peak. After a few months, some time series started to indicate the presence of a second wave. Consequently, the values of the ARE metric increased (i.e., worse projections) with respect to previous monitoring days. In order to circumvent this issue, the modelling approach was changed to deal with the sum of sub-regions forming a country, such as Brazil and its states; further details can be found in Chapter 5. After the "sum of sub-regions" strategy, a two-wave model was implemented and, few months later, the three-wave version was incorporated in the analysis.

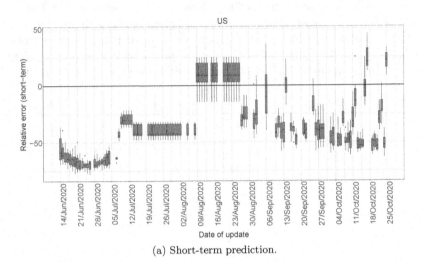

(a) Short-term prediction.

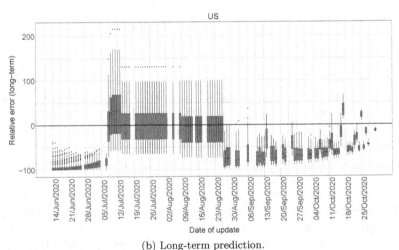

(b) Long-term prediction.

FIGURE 13.12: RE for both short- (a) and long-term (b) predictions of the number of new confirmed cases (COVID-19) in the United States. The analysis is based on data from 23 January 2020 to 31 October 2020. Data source: Center for Systems Science and Engineering at Johns Hopkins University.

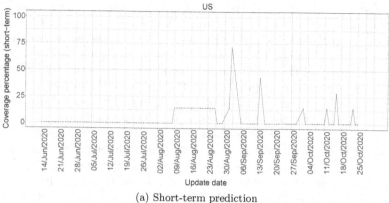

(a) Short-term prediction

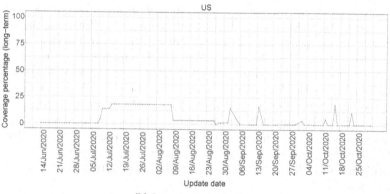

(b) Long-term prediction.

FIGURE 13.13: Coverage percentage for short- (a) and long-term (b) predictions related to the number of new confirmed cases (COVID-19) in the United States. The analysis is based on data from 23 January 2020 to 31 October 2020. The short-term prediction involves seven days ahead and the long-term includes all futures time points until the end of the pandemic. Data source: Center for Systems Science and Engineering at Johns Hopkins University.

Figure 13.11 shows that the estimated curve changed substantially throughout the period of study. Evidently, these differences are related not only to the arrival of new information, but also to the improvement of the adopted statistical model. Figure 13.12 illustrates the RE for both short- (Panel 13.12a) and long-term (Panel 13.12b) predictions; here the analysis involves data from 14 June until 31 October 2020. Short-term refers to the predictions seven days ahead and long-term accounts for projections related to several days ahead, after the last observation of the series. As a consequence of this distinction, note that the scale in the vertical axis is not the same for both panels. Given the modifications imposed in the model throughout the time, Figure 13.12 displays the RE boxplots in a configuration where different patterns can be seen separated in blocks.

Now, the analysis is focused on comparing the count series (grey dots) and the mean curves in Figure 13.11 with the results in Figure 13.12. According to the trajectory of the time series of confirmed cases in the United States, the second wave seems to start around 07 June 2020. When confronting this information with the pattern of boxplots in Figures 13.12a and 13.12b, one will find a justification for the unsatisfactory performance of the one-wave model. In other words, this version was not appropriate for the US series between 14 June and the beginning of July. Nonetheless, it is important to mention that the present analysis can only be done by using the full time series. The future observations were not available during the pandemic, therefore, the RE metric tends to increase the distance to the reference 0 immediately before the date when the model is replaced.

As previously mentioned, an improvement in the model was achieved by applying the technique of the sum of sub-regions. However, the second wave was already occurring for several sub-regions (states) within the country (United States). Therefore, the observed RE improvement was not strong enough in this case. The subsequent two-wave model specification adopted in the project for the United States series was able to fully account for the two-wave behaviour. This can be seen through the block of boxplots near 0 in the middle of Figure 13.12. In the United States, the third wave began in September and, consequently, the RE boxplots again increased their distance to 0. The three-wave model was then applied, but it was not able to substantially improve the RE. In this case, additional study is needed to reach a better result.

In terms of the prediction interval for the United States analysis, Figure 13.12a shows the coverage percentage related to the short- (Panel 13.13a) and long-term (Panel 13.13b) predictions. The results here lead to a clear conclusion that, even in the periods when the data was well fitted, the prediction intervals are too short to include the true counts. As a consequence, the coverage percentage is low for almost all days between 14 June to 31 October 2020.

A final aspect to be highlighted in Figure 13.13 is associated with the period from July to August 2020. Note that the coverage percentage is slightly higher between the mentioned months. Such performance may be a contribution of the sum of sub-regions model, as it aggregates the uncertainty from each sub-region leading to a wider interval in the bigger picture. This situation occurred mostly for the long-term analysis. The reader should refer to Chapter 4 for a discussion about strategies to obtain wider prediction intervals.

Bibliography

Almeida, F. M., Colosimo, E. A. and Mayrink, V. D. (2018) Prior specifications to handle the monotone likelihood problem in the Cox regression model. *Statistics and Its Interface*, **11**, 687–698.

— (2021) Firth adjusted score function for monotone likelihood in the mixture cure fraction model. *Lifetime Data Analysis*, **27**, 131–155.

Chowell, G., Luo, R., Sun, K., Roosa, K., Tariq, A. and Viboud, C. (2020) Real-time forecasting of epidemic trajectories using computational dynamic ensembles. *Epidemics*, **30**, 100379.

Chowell, G., Tariq, A. and Hyman, J. M. (2019) A novel sub-epidemic modeling framework for short-term forecasting epidemic waves. *BMC Medicine*, **17**.

Funk, S., Camacho, A., Kucharski, A. J., Lowe, R., Eggo, R. M. and Edmunds, W. J. (2019) Assessing the performance of real-time epidemic forecasts: A case study of Ebola in the Western Area region of Sierra Leone, 2014-15. *PLOS Computational Biology*, **15**.

Gelfand, A. and Smith, A. (1990) Sampling based approaches to calculating marginal densities. *Journal of the American Statistical Association*, **85**, 398–409.

Geman, S. and Geman, D. (1984) Stochastic relaxation, Gibbs distributions and the Bayesian restoration of images. In *IEEE Transactions Pattern Analysis and Machine Intelligence*, vol. 6, 721–741.

Hastings, W. (1970) Monte Carlo sampling using Markov chains and their applications. *Biometrika*, **57**, 97–109.

Hoffman, M. D. and Gelman, A. (2014) The No-U-Turn sampler: Adaptively setting path lengths in Hamiltonian Monte Carlo. *Journal of Machine Learning Research*, **15**, 1351–1381.

Hsieh, Y.-H. and Chen, C. W. S. (2009) Turning points, reproduction number, and impact of climatological events for multi-wave dengue outbreaks. *Tropical Medicine & International Health*, **14**, 628–638.

Hsieh, Y.-H. and Cheng, Y.-S. (2006) Real-time forecast of multiphase outbreak. *Emerging Infectious Diseases*, **12**, 122–127.

Hsieh, Y.-H., Fisman, D. N. and Wu, J. (2010) On epidemic modeling in real time: An application to the 2009 novel A (H1N1) influenza outbreak in Canada. *BMC Res Notes*, **3**.

Hsieh, Y.-H. and Ma, S. (2009) Intervention measures, turning point, and reproduction number for dengue, Singapore, 2005. *The American Journal of Tropical Medicine and Hygiene*, **80**, 66–71.

Kolassa, S. (2020) Why the "best" point forecast depends on the error or accuracy measure. *International Journal of Forecasting*, **36**, 208–211. M4 Competition.

Mayrink, V. D., Panaro, R. V. and Costa, M. A. (2021) Structural equation modeling with time dependence: an application comparing Brazilian energy distributors. *AStA Advances in Statistical Analysis*, **105**, 353–383. URLhttps://doi.org/10.1007/s10182-020-00377-2.

Metropolis, N., Rosenbluth, A., Teller, M. and Teller, E. (1953) Equations of state calculations by fast computing machines. *Journal of Chemistry and Physics*, **21**, 1087–1091.

Pell, B., Kuang, Y., Viboud, C. and Chowell, G. (2018) Using phenomenological models for forecasting the 2015 Ebola challenge. *Epidemics*, **22**, 62–70. The RAPIDD Ebola Forecasting Challenge.

Roosa, K., Lee, Y., Luo, R., Kirpich, A., Rothenberg, R., Hyman, J., Yan, P. and Chowell, G. (2020) Real-time forecasts of the COVID-19 epidemic in China from February 5th to February 24th, 2020. *Infectious Disease Modelling*, **5**, 256–263.

Schumacher, F. L., Ferreira, C. S., Prates, M. O., Lachos, A. and Lachos, V. H. (2021) A robust nonlinear mixed-effects model for COVID-19 deaths data. *Statistics and Its Interface*, **14**, 49–57.

Wang, X.-S., Wu, J. and Yang, Y. (2012) Richards model revisited: Validation by and application to infection dynamics. *Journal of Theoretical Biology*, **313**, 12–19.

14

Comparing predictions

Vinícius D. Mayrink
Universidade Federal de Minas Gerais, Brazil

Ana Julia A. Câmara
Universidade Federal de Minas Gerais, Brazil

Jonathan S. Matias
Universidade Federal de Minas Gerais, Brazil

Gabriel O. Assunção
Universidade Federal de Minas Gerais, Brazil

Juliana Freitas
Universidade Federal de Minas Gerais, Brazil

CONTENTS

The main aim of the present chapter is to show a comparison study confronting predictive results from the CovidLP project with those from the Institute for Health Metrics and Evaluation (IHME). The IHME is a health research centre, at the University of Washington, that developed a platform to show projections related to the COVID-19 pandemic for several regions around the world. The variety of regions, the data format, the easy access to downloads and the periodic updating of the series/projections were key aspects for choosing the IHME as the reference platform for the comparisons here. In addition, both CovidLP and IHME enable fast decisions, are focused on curve fitting to evaluate propagation, express uncertainty by providing prediction intervals and make modelling updates to accommodate new features. The study is focused on predictions involving 14 days ahead and it accounts for interval estimation and absolute relative errors calculated with respect to

DOI: 10.1201/9781003148883-14

the true observed counts. The reader should have in mind that the present analysis is not intended to discuss which platform is the best one in terms of predictions. Different cases displaying contrasting performances, for both platforms, were selected in this investigation.

14.1 The structure of the proposed comparison

The previous Chapters 12 and 13 were dedicated to discuss issues and obstacles related to the observations in the time series and the trajectory of the projected epidemiological curve, respectively. The main focus of the current chapter is to develop a comparative study confronting the outcomes from the CovidLP project with those from a similar platform providing results in the COVID-19 context. The chosen platform for this evaluation is the IHME, which is justified by two points: (i) it uses a modelling approach focused on curve fitting with prediction intervals and (ii) it releases and updates the data and projections throughout the pandemic. Both CovidLP and IHME show their periodically updated results for many regions in an online interactive application, where the methodology is well presented and downloads can be easily accessed by any user. According to the developers, the IHME COVID-19 project was created in response to requests from the University of Washington School of Medicine, United States hospital systems, and state governments interested in determining the moment at which the pandemic would overwhelm the health system capacity.

One major difference between the two platforms is the time period between updates. The CovidLP has a daily scheme for updating information, whereas the scheme adopted by the IHME is arbitrary, allowing 1 to 15 days between updates (the average is approximately 5 days). This distinction implies that the planned comparison to be discussed here should be done with respect to the projections obtained from using a time series observed until one of those days listed by the IHME updating scheme. Note that a large number of scenarios can be included in the proposed study due to the combination of several aspects such as region, day of the last observation in the time series, number of days to be predicted, and type of model being used for the projections. In order to simplify the presentation and avoid long and repeated discussions about graphs exhibiting similar behaviours, the scenarios investigated in this chapter will consider: a selection of a few interesting regions available from both platforms, no more than two final dates of the observed time series to be defined for both platforms and, finally, projections involving 14 days ahead. As previously explained in other chapters, the model adopted by the CovidLP project evolved during the pandemic. The main modification was related to the ability to handle more than one wave in the path of the series. As a result, it is also important to analyse how the estimates from the two-waves version improves with respect to the one-wave modelling.

A second difference between the platforms is the fact that IHME does not provide predictions for the number of confirmed cases. The time series of deaths is the focus of this platform, which is a choice potentially derived from the known higher accuracy of this type of response variable. In contrast, the CovidLP estimates the epidemiological curve for both confirmed cases and deaths. Given the mentioned distinction, the comparison "CovidLP × IHME" developed in the present chapter will only explore the serial death counts. The model comparison "one × two-waves", shown in the final section, is focused on the CovidLP and the series of confirmed cases.

The "14 days ahead" prediction is the standard configuration, assumed in this chapter, to explore results from any scenario. The incubation period for the COVID-19 virus is thought to extend to 14 days, therefore, this is one aspect justifying the number of predicted points. Obviously, larger periods could be considered, but in time series analysis, the predictive accuracy is known to reduce when increasing the distance to the last day with an observed value (Prado and West, 2010). In order to assess the performance of the two platforms under investigation, the absolute relative error (ARE) described in Chapter 13 is calculated here. This measurement is based on the absolute distance between each prediction and the corresponding true count. The reader must note that the mentioned true count to be predicted is not incorporated in the sequential data for which the model is fitted. Large values of the relative error indicate poor performance. The analysis will also consider 95% intervals for each prediction. These intervals describe the uncertainty of the projections. The corresponding amplitudes will depend on the chosen model, the distance to the last observation, the sample size (i.e., the length of the observed time series), and can also be influenced by the magnitude of the counts.

Researchers from many places around the world have dedicated their time to develop approaches for predicting the progression of cases/deaths related to the COVID-19. The list of epidemic forecasting methods ranges from basic curve fitting methodologies to machine-learning approaches; an example connected to the IHME research team is IHME COVID-19 Forecasting Team et al. (2020). As mentioned in previous chapters, the accurate estimation of the epidemiological curve, to identify when the outbreak will hit its peak, is extremely important for government planning and decision making. During the COVID-19 pandemic, many studies were dedicated to compare different prediction methods proposed in the literature; some examples are Achterberg et al. (2020), Appadu et al. (2020), Zeroual et al. (2020), Papastefanopoulos et al. (2020) and Friedman et al. (2020). The comparative analyses have greatly contributed to identify the advantages and disadvantages of each methodology. This type of study should always be done with caution to avoid judging a method as inappropriate, when in fact it can perform well depending on the behaviour of the data. The analysis in the present chapter is in line with this principle.

14.2   Analysis for Brazilian states

This section is dedicated to comparing predictions from CovidLP and IHME for some selected Brazilian states. As explained in the previous section, daily projections are not available from the IHME platform, therefore, two final dates of the time series (one is July 4 and the other is October 11 2020) are chosen from those listed with updated predictions in the IHME website. Again, it is important to reinforce to the reader that this analysis is based on estimates involving 14 days ahead. The investigation explores three main aspects: the trajectory of the epidemiological curve related to the COVID-19 death counts, 95% prediction intervals (PI) indicating the uncertainty related to the central trajectory, and finally, the ARE metric (see description in Chapter 13) summarising the distance to the true observations.

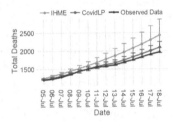
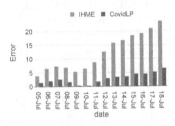

(a) Predictions and PI: 05–18 July 2020.   (b) ARE: 05–18 July 2020.

FIGURE 14.1: Predictions, 95% Intervals (PI), and Absolute Relative Error (ARE) for the number of COVID-19 confirmed deaths in Minas Gerais (Brazil). The time period is 05–18 July 2020. Data sources: Institute of Health Metrics and Evaluation and CovidLP project.

Figure 14.1 shows the results for Minas Gerais, which is located in the southeast region of the country and it has the second largest population ($\approx$ 21 million) among the Brazilian states. The analysis here is related to the period of 05–18 July 2020. As can be seen, the projections from the CovidLP project tend to be closer to the true observations. The 95% intervals from CovidLP are shorter and they all contain the true death counts. In contrast, all true values are found outside the 95% intervals obtained by the IHME modelling. The graph clearly indicates that the predictions from the IHME tend to overestimate the true trajectory in this situation. The absolute relative errors are expressed through a bar plot in Figure 14.1b. Note that the ARE is significantly lower for the CovidLP results related to all time points under investigation. In addition, this measurement has an overall increasing trend throughout the time for the IHME projections. This particular case is an example where CovidLP outperforms the IHME. However, this does not mean that other scenarios will indicate the same conclusion.

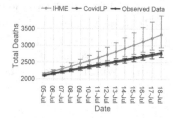

(a) Predictions and PI: 05–18 July 2020.

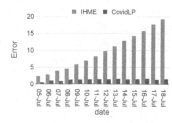

(b) ARE: 05–18 July 2020.

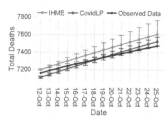

(c) Predictions and PI: 12–25 October 2020.

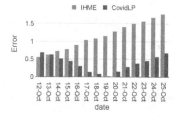

(d) ARE: 12–25 October 2020.

FIGURE 14.2: Predictions, 95% Intervals (PI), and Absolute Relative Error (ARE) for the number of COVID-19 confirmed deaths in Bahia (Brazil). The time periods are 05–18 July 2020 (Panels 14.2a and 14.2b) and 12–25 October 2020 (Panels 14.2c and 14.2d). Data sources: Institute of Health Metrics and Evaluation and CovidLP project.

Figure 14.2a presents the results corresponding to the state of Bahia, located in the northeast of Brazil (ranked 4th in population size). In this analysis the comparison "CovidLP × IHME" is done based on two time periods. Panel 14.2a indicates a result quite similar to the one observed for Minas Gerais. In other words, the CovidLP provides better predictions than the IHME platform. The true values are captured by all 95% intervals from CovidLP. Intervals from IHME are generally higher than the true counts, except the first three (05–07 July) with the observed value near the lower limit. Panel 14.2b displays the magnitudes of the ARE for both platforms in the period of 05–18 July 2020. This result reflects the last discussion about the proximity between the predictions and the true counts. In summary, for any given time point, the ARE from CovidLP is lower than the one from IHME. In addition, a monotonically increasing pattern across the days is observed for the IHME case.

Figure 14.2c presents complementary results, related to Bahia, to evaluate the platforms in another time period (12–25 October). In this scenario, note that the predictive performance of CovidLP is still good, with 95% intervals including the true count for almost all days (except 12–14 October). The IHME intervals were unable to capture the true values. The projections in light grey clearly overestimate the true number of deaths. Panel 14.2d indicates once again the monotonically increasing pattern of ARE measurements obtained via IHME. The same trend is not detected for the CovidLP; here the ARE decreases until 19 October and then increases. As a final comment, 12–13 October are the only days for which the ARE from CovidLP is not smaller than the one from IHME.

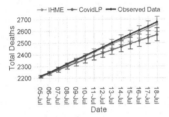

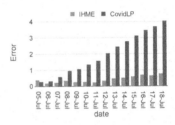

(a) Predictions and PI: 05–18 July 2020.   (b) ARE: 05–18 July 2020.

FIGURE 14.3: Predictions, 95% Intervals (PI), and Absolute Relative Error (ARE) for the number of COVID-19 confirmed deaths in Maranhão (Brazil). The time period is 05–18 July 2020. Data sources: Institute of Health Metrics and Evaluation and CovidLP project.

The previous illustrations involving Minas Gerais and Bahia indicated a strong performance of the CovidLP projections. As already commented in this chapter, one should not assume that this behaviour can be extended to all other regions and time periods. Figure 14.3a investigates the data related to the Brazilian state of Maranhão (northeast region, ≈ 7 million inhabitants). The time period under analysis is 05–18 July 2020. As can be seen, this example shows better performance of the IHME with respect to the CovidLP. Note that the light grey points are closer to the black line. Consequently, all 95% intervals from IHME contain the true counts. The CovidLP intervals were only able to capture the true values, near the upper limit, for the period of 05–11 July 2020. Panel 14.3b shows that a monotonically increasing trend is now observed for the ARE from the CovidLP. The pattern associated with the IHME is not monotone and it indicates lower light grey bars for almost all time points (except 05 July).

Figure 14.4 presents another case where the IHME has better predictive performance than the CovidLP. The analysis is related to the state of Rio Grande do Sul, ranked with the 6th largest population (≈ 11 million) and located in the south region of Brazil. Panel 14.4a corresponds to the period 05–18 July 2020. Note that the estimates provided by the IHME are closer to the true counts, consequently, all 95% intervals in light grey include the black dots. The CovidLP estimates are clearly below the black line. In addition, the 95% intervals (dark grey segments) do not incorporate the true value or the value is near the upper limit. This analysis suggests a configuration where predictions are underestimated via CovidLP. Panel 14.4b is in accordance with the previous discussion. An increasing pattern is clear for the dark grey bars representing the ARE from CovidLP. Most of them are higher than the light grey bars from IHME.

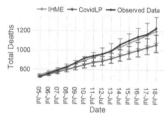

(a) Predictions and PI: 05–18 July 2020.

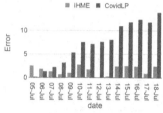

(b) ARE: 05–18 July 2020.

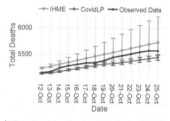

(c) Predictions and PI: 12–25 October 2020.

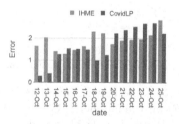

(d) ARE: 12–25 October 2020.

FIGURE 14.4: Predictions, 95% Intervals (PI), and Absolute Relative Error (ARE) for the number of COVID-19 confirmed deaths in Rio Grande do Sul (Brazil). The time periods are 05–18 July 2020 (Panels 14.4a and 14.4b) and 12–25 October 2020 (Panels 14.4c and 14.4d). Data sources: Institute of Health Metrics and Evaluation and CovidLP project.

A different result is detected for predictions using the time series of Rio Grande do Sul until 11 October 2020. Figure 14.4c shows the projections for the period 12–25 October 2020. As can be seen, the black dots representing the true counts are located between the estimates from both platforms. The 95% intervals from IHME have larger amplitude than those from CovidLP. As a consequence, almost all intervals from IHME include the true counts (except in 12–14 October). The CovidLP does not have the same performance in terms of intervals capturing the true values. The ARE bars in Panel 14.4d do not indicate the large magnitudes and increasing trend observed for the CovidLP in Panel 14.4b. In the present scenario, it is difficult to judge which platform has the overall best performance in terms of ARE; 50% of the dark grey bars are higher than the light grey ones.

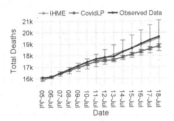

(a) Predictions and PI: São Paulo.

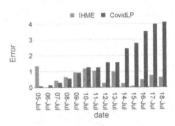

(b) ARE: São Paulo.

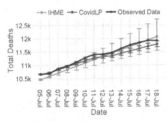

(c) Predictions and PI: Rio de Janeiro.

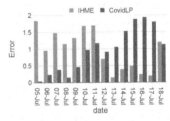

(d) ARE: Rio de Janeiro.

FIGURE 14.5: Predictions, 95% Intervals (PI), and Absolute Relative Error (ARE) for the number of COVID-19 confirmed deaths in the Brazilian states of São Paulo (Panels 14.5a and 14.5b) and Rio de Janeiro (Panels 14.5c and 14.5d). The time period is 05–18 July 2020. Data sources: Institute of Health Metrics and Evaluation and CovidLP project.

Figure 14.5 exhibits the predictions and the ARE measurements for São Paulo and Rio de Janeiro. These are the first (≈ 46 million) and third (≈ 17 million) most populous states in Brazil, respectively. Here, the period under study is 05–18 July 2020. Panel 14.5a indicates that IHME has in general greater proximity to the true counts in black. The CovidLP predicts well the number of deaths in 05–09 July, but the distance to the true value increases after this period. Panel 14.5b shows the ARE bars for São Paulo. An overall monotonically increasing pattern is clearly seen for the CovidLP errors. The graph in Panel 14.5c provides the projections for Rio de Janeiro. In this case, CovidLP has better proximity to the black line in the initial period (05–11

July); IHME indicates lower distances in 12-17 July. This illustration shows again, for both states, that the 95% intervals from CovidLP are shorter than those from IHME; this aspect is explained by the different models and inference approaches considered by the platforms. Panel 14.5d displays the ARE measurements related to Rio de Janeiro. One can clearly see the higher light grey (IHME) bars for 05–11 July and the higher dark grey (CovidLP) bars for 12–17 July.

TABLE 14.1: Summary statistics (mean and median) of the ARE metrics obtained in the projections for each Brazilian state.

	CovidLP			
	05–18 July		12–25 October	
State	Mean	Median	Mean	Median
Minas Gerais	2.8540	2.7464	0.9874	1.0913
Bahia	1.2684	1.3619	1.1459	0.4099
Maranhão	1.9687	1.7982	1.3863	1.3343
Rio Grande do Sul	7.3566	7.5483	1.6725	1.5198
Rio de Janeiro	0.9718	1.0104	1.5868	1.5273

	IHME			
	05–18 July		12–25 October	
State	Mean	Median	Mean	Median
Minas Gerais	12.3360	10.7130	0.6149	0.7664
Bahia	9.6780	8.9780	1.1459	1.1193
Maranhão	0.4197	0.3631	0.2321	0.2370
Rio Grande do Sul	1.6153	1.7587	1.8910	1.9120
Rio de Janeiro	0.9621	1.0519	0.1462	0.1425

The investigation exploring the Brazilian states is now completed. In brief, the discussions developed here indicate that the predictive performance of the studied platforms depends on the data set. Contrasting conclusions can be obtained for different regions and for the same region in different time periods. Table 14.1 presents a final summary of the ARE measurements obtained in the projections for each Brazilian state, which were previously investigated in the different graphs of this section. All main conclusions discussed in graphs are reflected in Table 14.1. Additional prediction results for other states are available in the supplementary material of the book at www.github.com/CovidLP/book. In the next section, the illustrations will evaluate data from a selection of countries.

14.3   Analysis for countries

The "CovidLP × IHME" comparison is developed in the present section for time series related to the following countries: Colombia, Mexico, Turkey, Argentina, Brazil, and the United States. The time periods (with 14 days) for predictions are 15–28 July and 16–29 October 2020. These two periods were chosen according to the list of days having updated projections in the IHME website; recall that IHME does not update their results in a daily basis. The six mentioned countries were selected due to the interesting scenarios observed in their evaluation. The analyses for Colombia, Mexico, Turkey and Argentina will discuss the differences between the estimates obtained from the two time periods under investigation. Brazil and the United States are specific cases to be explored in the end of this section. In general, the discussions will evaluate the proximity to the true counts and summarise this information through ARE measurements.

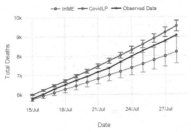

(a) Predictions and PI: 15–28 July 2020.

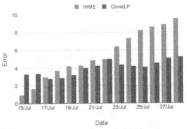

(b) ARE: 15–28 July 2020.

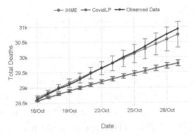

(c) Predictions and PI: 16–29 October 2020.

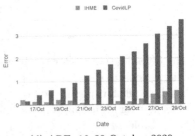

(d) ARE: 16–29 October 2020.

FIGURE 14.6: Predictions, 95% Intervals (PI), and Absolute Relative Error (ARE) for the number of COVID-19 confirmed deaths in Colombia. The time periods are 15–28 July 2020 (Panels 14.6a and 14.6b) and 16–29 October 2020 (Panels 14.6c and 14.6d). Data sources: Institute of Health Metrics and Evaluation and CovidLP project.

Figure 14.6 presents the results for Colombia. Panel 14.6a is related to the period 15–28 July, and it shows a configuration where the true counts are positioned between the predictions from both platforms. The CovidLP tends to overestimate the death counts, whereas the IHME underestimates these quantities. Similar to what was observed in the previous section, the 95% intervals from the CovidLP are shorter than those from IHME. A visual inspection of the graph indicates that, in general, the dark grey line (CovidLP) seems closer to the black line (true values) than the light grey line (IHME). Panel 14.6b confirms this interpretation, since the majority of light grey bars (IHME) are higher than the dark grey ones (CovidLP). A monotonically increasing trend is observed for the ARE measurements obtained from IHME predictions. This particular illustration indicates the better performance of the CovidLP platform.

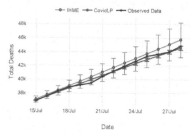

(a) Predictions and PI: 15–28 July 2020.

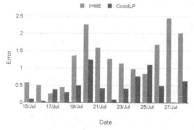

(b) ARE: 15–28 July 2020.

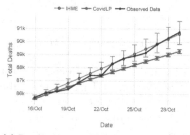

(c) Predictions and PI: 16–29 October 2020.

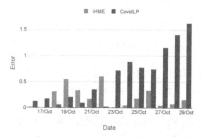

(d) ARE: 16–29 October 2020.

FIGURE 14.7: Predictions, 95% Intervals (PI), and Absolute Relative Error (ARE) for the number of COVID-19 confirmed deaths in Mexico. The time periods are 15–28 July 2020 (Panels 14.7a and 14.7b) and 16–29 October 2020 (Panels 14.7c and 14.7d). Data sources: Institute of Health Metrics and Evaluation and CovidLP project.

The graph in Panel 14.6c is related to the second time period (16–29 October 2020). As can be seen, the behaviour of the predictions for Colombia changes in this scenario. Note that the true values are closer to the IHME trajectory; the corresponding 95% intervals (light grey segments) include the true counts. Here, the CovidLP underestimates the number of deaths. The ARE bars, shown in Panel 14.6d, exhibit a clear monotonically increasing pattern for the CovidLP measurements. The light grey bars (IHME) are substantially lower than those in dark grey. This illustration reinforces the idea that a good predictive performance for one time period does not imply that the same platform will perform well in another period for the same region.

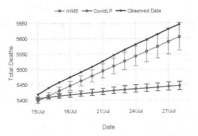

(a) Predictions and PI: 15–28 July 2020.

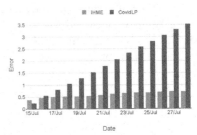

(b) ARE: 15–28 July 2020.

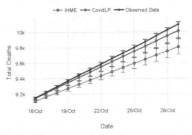

(c) Predictions and PI: 16–29 October 2020.

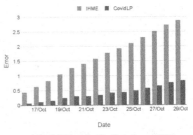

(d) ARE: 16–29 October 2020.

FIGURE 14.8: Predictions, 95% Intervals (PI), and Absolute Relative Error (ARE) for the number of COVID-19 confirmed deaths in Turkey. The time periods are 15–28 July 2020 (Panels 14.8a and 14.8b) and 16–29 October 2020 (Panels 14.8c and 14.8d). Data sources: Institute of Health Metrics and Evaluation and CovidLP project.

Figure 14.7 evaluates the results for Mexico. Panel 14.7a indicates that the wide 95% intervals from IHME incorporate all true counts and also the shorter intervals from the CovidLP. The visual inspection of the distance between grey and black points seems to indicate an overall better performance of the CovidLP. This aspect is confirmed by the bar plot in Panel 14.7b. Note that 12 out of 14 days have lower ARE measurements for the CovidLP platform (17 and 25 July are the only dates favouring the IHME). Panel 14.7c displays the predictions for Mexico in 16–29 October. In this example, the IHME outperforms the CovidLP predictions in the second half of the series

(23–29 October); the black line is clearly closer to the light grey estimates. The ARE bars shown in Panel 14.7d reflect this interpretation. As can be seen, the dark grey bars (CovidLP) are higher than the light grey ones for 23–29 October. In summary, 10 out of 14 days have lower ARE measurements for the IHME platform in this particular analysis.

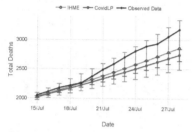

(a) Predictions and PI: 15–28 July 2020.

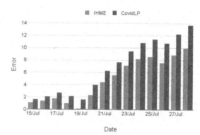

(b) ARE: 15–28 July 2020.

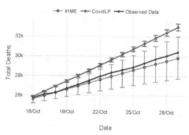

(c) Predictions and PI: 16–29 October 2020.

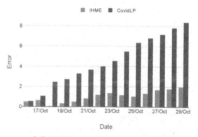

(d) ARE: 16–29 October 2020.

FIGURE 14.9: Predictions, 95% Intervals (PI), and Absolute Relative Error (ARE) for the number of COVID-19 confirmed deaths in Argentina. The time periods are 15–28 July 2020 (Panels 14.9a and 14.9b) and 16–29 October 2020 (Panels 14.9c and 14.9d). Data sources: Institute of Health Metrics and Evaluation and CovidLP project.

Figure 14.8 shows the predictions for Turkey. This is an interesting example where CovidLP has poor performance in 15–28 July, but it shows better results for 16–29 October 2020. Panel 14.8a presents the path of the predictions obtained for the period in July. Both platforms underestimate the true counts in black, but the visual analysis clearly indicates greater proximity to the truth for the IHME projections. In line with this conclusion, Panel 14.8b exhibits a monotonically increasing pattern of ARE measurements related to the CovidLP. The ARE values from IHME are lower than those from CovidLP for all days, except 15 July. Panels 14.8c and 14.8d, corresponding to 16–29 October, determine an inverted conclusion. As can be seen, both platforms underestimate the true counts, but now the CovidLP predictions are closer to the black line. All 95% intervals in dark grey (CovidLP) include the black dots. On the other hand, the true values are outside the range of the light grey intervals (IHME). The bar plot in Panel 14.8d, representing the magnitude of the ARE, suggests an increasing pattern for both platforms, however, all light grey bars (IHME) are substantially higher than those in dark grey (CovidLP).

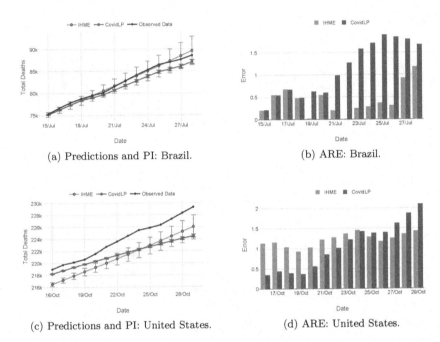

(a) Predictions and PI: Brazil.

(b) ARE: Brazil.

(c) Predictions and PI: United States.

(d) ARE: United States.

FIGURE 14.10: Predictions, 95% Intervals (PI), and Absolute Relative Error (ARE) for the number of COVID-19 confirmed deaths in Brazil (Panels 14.10a and 14.10b, period of 15–28 July 2020) and the United States (Panels 14.10c and 14.10d, period of 16–29 October 2020). Data sources: Institute of Health Metrics and Evaluation and CovidLP project.

Figure 14.9 presents two scenarios related to Argentina. Again, the analysis is based on two periods: 15–28 July and 16–29 October 2020. Panel 14.9a shows the predictions and 95% intervals for July. As can be seen, both platforms have estimates near the true values for the first five days. This behaviour changes for the days after 20 July. Note that a fast increase is observed for the true counts and the predictions remain in a lower level leading to underestimation. This visual analysis is confirmed by the ARE bars in Panel 14.9b. The measurements related to CovidLP are slightly higher than those from IHME. Panels 14.9c and 14.9d exhibit the results for October. In this case, the CovidLP overestimates the true counts and indicates a monotonically increasing pattern of ARE measurements.

Figure 14.10 contains the results for Brazil (15–28 July 2020) and the United States (16–29 October 2020). During the COVID-19 pandemic, these two countries reported large daily numbers of confirmed cases and deaths, therefore, they are chosen for the last comparison in the current section. In order to show configurations that do not indicate a single platform with low ARE for all time points, the period of predictions is not the same for both countries. Panels 14.10a and 14.10b are related to Brazil. The main conclusion can be obtained by looking at the ARE bars. Note that both platforms have similar performance for the initial dates (15–18 and 20 July). The distance to the true count is significantly larger for the CovidLP predictions after 20 July. Panels 14.10c and 14.10d display the results for the United States. The visual inspection of the estimates suggests that both platforms underestimate the true number of deaths. The CovidLP predictions have higher proximity to the black line in the first nine days. This conclusion is confirmed by the bar plot of ARE measurements. Note that the dark grey bars (CovidLP) are lower than the light grey ones (IHME) for 16–24 October.

The reader is reminded that the statistical model adopted by the CovidLP project evolved throughout the pandemic. Strategies were implemented to improve the predictive accuracy of the platform. One of these strategies was to change the model to allow a more appropriate fit with respect to the presence of a two-waves scenario in the series. This alteration improved the predictions obtained from the CovidLP project. The next section is dedicated to compare the projections from CovidLP before and after the mentioned modelling modification. Additional prediction results related to other countries are available in the supplementary material of the book at `www.github.com/CovidLP/book`.

14.4 Improvements from a two-waves modelling

The purpose of this section is to evaluate the impact of modifying the statistical model, assuming one-wave, to a setting where the epidemiological curve can have two increasing periods (two-waves). This is an important alteration implemented to handle the time series related to countries or states in which the two-waves feature was visually clear at some point during the pandemic. Naturally, the one-wave version was perfectly appropriate to fit all time series in the initial part of the pandemic, but this aspect changed after some months. The dates in which the second wave was detected is not the same for all regions, therefore, monitoring the path of all time series was important to identify the right moment for the alteration. The reader is reminded that a discussion about detecting the second wave can be found in Chapter 12.

Only the predictions (from CovidLP) for the number of confirmed cases are investigated here. The reason for using this response is the fact that the presence of a second wave is in general more evident in the trajectory of the series for the confirmed cases. In short, the analysis is developed with the following steps: (i) identify the day t in which the CovidLP research team decided to apply the two-waves model, (ii) obtain the predictions from the two-waves model using the data observed until day t, (iii) obtain the predictions from the one-wave model based on the data observed until day $t-1$. The estimates from steps (ii) and (iii) are compared to investigate the proximity to the true counts.

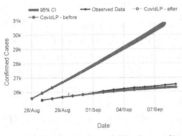
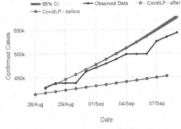

(a) Predictions and PI: Australia.   (b) Predictions and PI: Spain.

FIGURE 14.11: Predictions and 95% Intervals (PI) corresponding to the models assuming the one-wave (before model alteration) and two-waves (after model alteration) settings. Panels 14.11a and 14.11b evaluate Australia and Spain, respectively. The one-wave model projections are based on the data observed until 26 August (two-waves model applied on 27 August 2020).

Figure 14.11 presents the results for Australia and Spain. The two-waves model was established for both countries in 27 August 2020. Panel 14.11a indicates better predictive accuracy when fitting the two-waves modelling. As can be seen, the one-wave model overestimates the true values, whereas the two-waves alternative provides estimates quite close to the true path. The graph in Panel 14.11b shows a similar improvement. In this case, the one-wave model underestimates the true counts and the two-waves version has greater proximity to the black line. In summary, the present analysis indicates that the adaptation to deal with the two-waves setting determines a substantial improvement in the predictive accuracy of the CovidLP platform.

Bibliography

Achterberg, M. A., Prasse, B., Ma, L., Trajanovski, S., Kitsak, M. and Van Mieghem, P. (2020) Comparing the accuracy of several network-based COVID-19 prediction algorithms. *International Journal of Forecasting*.

Appadu, A., Kelil, A. and Tijani, Y. (2020) Comparison of some forecasting methods for COVID-19. *Alexandria Engineering Journal.*

Friedman, J., Liu, P., Troeger, C. E., Carter, A., Reiner, R. C., Barber, R. M., Collins, J., Lim, S. S., Pigott, D. M., Vos, T., Hay, S. I., Murray, C. J. L. and Gakidou, E. (2020) Predictive performance of international COVID-19 mortality forecasting models. *medRxiv.*

IHME COVID-19 Forecasting Team, Reiner, R. C., Barber, R. M. and et al. (2020) Modeling COVID-19 scenarios for the United States. *Nature Medicine.*

Papastefanopoulos, V., Linardatos, P. and Kotsiantis, S. (2020) COVID-19: A comparison of time series methods to forecast percentage of active cases per population. *Applied Sciences,* **10**, 3880.

Prado, R. and West, M. (2010) *Time Series: Modeling, Computation, and Inference.* New York: Chapman and Hall/CRC, 1st edn.

Zeroual, A., Harrou, F., Dairi, A. and Sun, Y. (2020) Deep learning methods for forecasting COVID-19 time-series data: A comparative study. *Chaos, Solitons & Fractals,* **140**, 110121.

Part VI

Software

15

PandemicLP package: Basic functionalities

Marcos O. Prates
Universidade Federal de Minas Gerais, Brazil

Guido A. Moreira
Universidade Federal de Minas Gerais, Brazil

Marta Cristina C. Bianchi
Universidade Federal de Minas Gerais, Brazil

Débora F. Magalhães
Universidade Federal de Minas Gerais, Brazil

Thais P. Menezes
University College Dublin, Ireland

CONTENTS

DOI: 10.1201/9781003148883-15

In this chapter we introduce the `PandemicLP` package, available in the statistical software R (R Core Team, 2020). This package aims to provide a straightforward tool for applying the Bayesian approach proposed by the CovidLP group for epidemiological data. Specifically, it offers functions to fit the models discussed in Chapters 3 and 5. The package implementation objectives were threefold: 1) provide control and versioning for the codes of the CovidLP project; 2) facilitate the automation of the code for the project app; and 3) allow a simple interface for beginners and advanced R users to run Bayesian inference in the models used by the CovidLP group. With this in mind, we present now the basic functionalities of the package, illustrating them with COVID-19 data modelling examples.

15.1   Introduction

To provide simple, reliable and adequate maintenance of the implementation codes for the CovidLP app, the team felt the need to establish a procedure that allowed tracking of these codes for the different methodologies over time. The `PandemicLP` package, available in the statistical software R (R Core Team, 2020), was created to establish version control and facilitate the use of our methodology by interested parties that wished to analyse different areas not contemplated by our application. This chapter was written for the `PandemicLP` package version 0.2.1 in software R version 4.0.3.

Most R packages available in the Comprehensive R Archive Network (CRAN) and GitHub, concerning specifically the COVID-19 pandemic, were primarily focused on providing reliable and up-to-date data for different regions, for example, the packages `COVID19` (Guidotti and Ardia, 2020), `covid19br` (Demarqui, 2020), and `coronavirus` (Krispin and Byrnes, 2020). Meanwhile, other packages such as `covid19.analytics` (Ponce and Sandhel, 2020) go beyond importing data and also provide calculations for different measures as well as plotting capabilities. From an epidemiological perspective, there are more general R packages such as the `EpiModel` (Jenness et al., 2018) and a series of packages created by the R Epidemics Consortium (RECON).

As will be shown and discussed, our package offers new features that are advantageous in comparison to the previously cited ones: 1) the `PandemicLP` package provides the capabilities of extracting data (for the COVID-19 pandemic), modelling, predicting and plotting results all in one package; 2) beginner users can easily utilise this package to create predictions based on MCMC methods at the same time that it allows advanced users to tweak different parameters at will; 3) the package produces long-term predictions that enable the estimate of the peak and end of the pandemic/epidemic to better guide decision making; and 4) although the `PandemicLP` package was created to model the COVID-19 pandemic, its models are not restricted to this disease and it can be applied in any other epidemic, as long as the input data is formatted accordingly.

15.2 Installation

The `PandemicLP` package is available on CRAN, so the installation process is the usual one used for most R packages (Listing 15.1).

```
install.packages("PandemicLP")
```

Listing 15.1: Installation of the `PandemicLP` package from CRAN

In the package, there are two example vignettes available that can be accessed through the commands `vignette("PandemicLP")` and `vignette("PandemicLP_SumRegions")`.

It is important to highlight that the package relies on and uses functions from some other packages which will be installed automatically if they are not installed yet. The main one is the `rstan` package (Stan Development Team, 2020), which in the current version depends on the V8 JavaScript libraries. If the installation of `rstan` fails in your Unix-based system, you might need to install it beforehand. On Windows, to fix this, the user might need to install the V8 package (Ooms, 2020) from the CRAN binaries by running `install.packages("V8")`. Installation on Mac requires at least the 3.6.2 macOS version.

In case the user wants to see the code behind each function, he or she can check it on the GitHub repository (`github.com/CovidLP/PandemicLP`). Besides that, if for some reason the usual installation process fails, the user can choose to install the package from the GitHub repository or using the binaries available there.

15.2.1 Installing from the GitHub repository

To install the package from the GitHub repository, it is necessary to install the `devtools` package (Wickham et al., 2020) with the function `install.packages("devtools")`. When the installation process is complete, the user must run the code presented in Listing 15.2 to install the `PandemicLP` package. The argument `build_vignettes = TRUE` will make the installation take longer, but it will make the examples vignettes available for viewing.

```
devtools::install_GitHub("CovidLP/PandemicLP",
                         build_vignettes = TRUE)
```

Listing 15.2: Installation of the `PandemicLP` package from the GitHub repository

15.3   Functionalities

The `PandemicLP` package version 0.2.1 has six main functions or methods for the statistical analysis of epidemiological data. For the COVID-19 data, it provides tools for data extraction through the `load_covid` function and allows plotting the observed data series with the generic function `plot`, that relies on the method `plot.pandemicData`. The model fitting of the package is obtained by the function `pandemic_model` that returns an object of the class `pandemicEstimated`. This function can be utilised for any epidemiological data as long as the data is formatted adequately. The generic function `posterior_predict`, using the implemented method `posterior_predict.pandemicEstimated`, is responsible for creating a sample of the predicted values of a given `pandemicEstimated` object. Next, a summary function, `pandemic_stats`, calculates relevant statistics of the `posterior_predict` output such as the 95% credible interval for short-term and long-term predictions, information about the estimated peak and end of the pandemic. Additionally, the `plot` function, by the method `plot.pandemicPredicted`, also allows the user to visualise the prediction results with the observed data values and its relevant statistics in a graph format. The general workflow is illustrated in Figure 15.1.

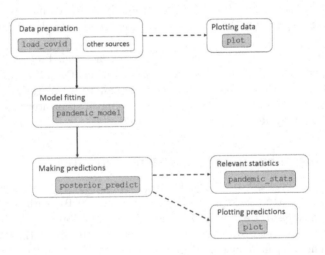

FIGURE 15.1: The `PandemicLP` package workflow.

Details on each function or method are provided in the following subsections, including examples of its use with the COVID-19 pandemic data in different locations.

15.3.1 COVID-19 data extraction and loading: `load_covid`

Although the models in the `PandemicLP` package were created to work with any epidemiological data, the COVID-19 pandemic was the initial motivation for producing this package. Therefore, the function `load_covid`, as the name suggests, loads COVID-19 data from reliable online repositories into R. This function formats the data to be used directly into the model fitting function `pandemic_model` of this package or to be plotted using the function `plot`.

The repository chosen for Brazil and its states was the *Observatório COVID-19 BR* (`github.com/covid19br/covid19br.github.io`), whereas for the remaining countries, the Center for Systems Science and Engineering (CSSE) at Johns Hopkins University was utilised (`github.com/CSSEGISandData/COVID-19`). For more information on how the repositories were chosen, see Section 9.1.

The `load_covid` function contains three arguments: `country_name`, `state_name` and `last_date`. The user must specify the country of interest through the argument `country_name`. To make it easier for the user to know the available countries in the CSSE database where data can be loaded, the function `country_list` was created. At the time that this book was written, the argument `state_name` was only available for states within Brazil. The function `state_list` is equivalent to the `country_list` function and returns the state abbreviations to be used as input and their respective state names. Lastly, the argument `last_date` allows the user to determine the latest date of the data series to select. The dates must be in the '`YYYY-MM-DD`' format. This input argument could be useful when comparing the predictions made by a certain model with the actual observations.

The resulting output object of function `load_covid` is a list with three items:

- `data`: a data frame with the number of cumulative cases, new cases, cumulative deaths and new deaths associated with COVID-19 for each date, up to the `last_date` in the specified region.

- `name`: string with the country name (and state name, if available).

- `population`: numeric object that contains the population size of the given region.

15.3.2 Visualising the data: `plot.pandemicData`

For graphical visualisation of the data, the package has the method `plot.pandemicData` implemented, which can be used through the function `plot`. Using the data resulting from the `load_covid` function or formatted as will be explained in Section 15.6, the method generates a plot with the daily cases and deaths or the cumulative results. An available option, `cases = "both"`, plots both plots simultaneously. The cases and deaths series are put together in the same graph, but each series can be seen individually by double clicking on the legend. This is a `plotly` resource, which generates an interactive graph in a way that a double-click on the legend isolates one trace. All series are plotted back when clicked again.

The `plot.pandemicData` method has the following arguments:

- `x`: an object of class `pandemicData` created by the function `load_covid`.

- `cases`: a string which indicates whether plots of new cases, cumulative cases, or both should be generated. The argument must be a string being either 'new', 'cumulative' or 'both'. The default of the method is to plot only the new cases.

- `color`: a logical variable indicating whether the plot should be coloured or in greyscale. The argument must be either TRUE or FALSE. The default of the method is to plot in colours.

The plots can be assigned to a variable which will be a list containing two objects: `new` and `cumulative`. The "new" can be accessed using `variable$new` and it is the plot for daily new confirmed cases and daily new deaths. The "cumulative" object is the plot for daily cumulative cases and daily cumulative deaths and can be shown using `variable$cumulative`. If any of them did not get plotted due to lack of prediction or due to the cases argument, its value will return NULL.

15.3.3   Model fitting: `pandemic_model`

The function `pandemic_model` fits a growth curve model to count data. In the COVID-19 example it can be the new cases of either confirmed or deaths. In version 0.2.1 of the package, the generalised logistic model for one or two waves is available. For details about these models, see Chapter 3 and Chapter 5. A seasonal effect can be included in the one-wave model according to Equation (5.2). In this version, the data is only modelled by the Poisson distribution. For a brief description of these models, see the package documentation, `?models`.

In the Bayesian framework, this function produces a posterior sample for every parameter of the model, where the point estimate for every parameter is usually the respective posterior median (or mean). These posterior samples are generated by the sampler of the Bayesian open-source software `Stan` (`mc-stan.org`) through its R interface, the `rstan` package version 2.21.1. The choice of the `Stan` software and its specifics are discussed in Section 10.1.

The main objective of this subsection is to present the functionalities of the `pandemic_model` function when keeping the default settings used in the app of the CovidLP project. As presented in Chapter 10, this is done by the command `CovidLPconfig=TRUE`. It is worth emphasising that the use of `CovidLPconfig=TRUE` may not guarantee chain convergence for different data analysis. For an advanced user, Chapter 16 is intended to explore these and other settings of the `Stan` sampler.

The package documentation for the `pandemic_model` presents a description of all input arguments. While maintaining the default setting `CovidLPconfig=TRUE`, it suffices to consider the following input arguments:

- `Y`: an object of class `pandemicData` created by the `load_covid` function or a list providing other epidemiological data for the model. However, in the examples provided in this section, `Y` was considered to be the object returned by function `load_covid`. Section 15.6 exemplifies how data from other sources (outside from `load_covid`) should be formatted.

- `case_type`: a string providing the type of cases of interest in modelling the epidemic. Current options are `"confirmed"` for confirmed cases or `"deaths"` for deaths. The default is `"confirmed"`.

- `seasonal_effect`: string vector indicating the days of the week in which seasonal effect was observed. The vector can contain the full weekday name (Sunday to Saturday) or the first 3 letters, up to a maximum of three weekdays. In version 0.2.1 of the package this argument can only be used for `n_waves = 1`. The default is `NULL`.

- `n_waves`: integer 1 or 2. This argument indicates the number of waves to be adjusted by the mean curve. The default is 1.

The output of a model fitted by the function `pandemic_model` is a list. This list contains the element `fit`, the output of the `Stan` sampler, among other elements of interest when fitting the model, such as `config.inputs` that shows settings used by the `Stan` sampler and the restrictions on the model parameters (if there are any). The `print` and `summary` methods present point estimates (mean and median) and some quantiles of the posterior sample for each parameter in the model, as well as its prior specifications. Also, these methods provide some diagnostic statistics for efficiency and chain convergence (see Chapter 16).

15.3.4 Predictive distribution: `posterior_predict.pandemicEstimated`

The package `rstantools` (Gabry et al., 2020) has a generic function for S3 methods called `posterior_predict`. Packages that use `rstan` code are encouraged to use this function for predictions. This has been implemented with the use of the method `posterior_predict.pandemicEstimated`, that is, an S3 method for the class `pandemicEstimated`.

This method provides calculations of the predictive distribution, which is defined in Equation (6.7). The prediction is done not only in terms of future counts y_t, but also the mean values $\mu(t)$. The sample of the future values of $\mu(t)$

are useful for many calculations performed in the function `pandemic_stats`, described in Section 15.3.5. The theoretical background for the procedures performed are described in Chapter 6.

In summary, the function reads which is the fitted model from the input object and applies an adequate prediction procedure. There are some fail-safe measures implemented so that absurd values are not considered in the prediction. Specifically, samples from the posterior which result in a cumulative prediction of values greater than the location's population are excluded. The indexes of the removed samples are included in the function output as `errors`.

There are two inputs for the function, namely `horizonLong` and `horizonShort`, where both need to be positive integers and the former must be larger than the latter. They control how far into the future the predictions are to be made. It is important to note that this function exclusively provides the predictive distribution for every time point in the `horizonLong` future. For this reason, summaries of the predictive distribution are calculated in a separate function, described in the next section.

The output of this function is an S3 object of the class `pandemicPredicted` and it is a list containing the following elements.

- `predictive_Long`: an $M \times$`horizonLong`-dimensional matrix, where M is the predictive sample size inherited from the `pandemic_model` function call and `horizonLong` is the input of the function. Each column is a sample of the predictive distribution of each point to the future. The prediction is towards the daily new confirmed or deaths cases, as defined in the `pandemic_model` call.

- `predictive_Short`: an $M \times$`horizonShort`-dimensional matrix, where M is the predictive sample size inherited from the `pandemic_model` function call and `horizonShort` is the input of the function. Each column is a sample of the predictive distribution of each point to the future. The prediction is towards the daily cumulative confirmed or deaths cases, as defined in the `pandemic_model` call.

- `data`: the raw data, as passed on to the `pandemic_model` function in the form of a `data.frame`.

- `location`: a string describing the location relative to the data. Inherited from the `pandemic_model` function call.

- `cases_type`: a string containing either "confirmed" or "deaths". Inherited from the `pandemic_model` function call.

- `pastMu`: an $M \times n$-dimensional `data.frame` containing a posterior sample for the mean function of the observed values, where M is the predictive sample size inherited from the `pandemic_model` function call and n is the number of observed counts in the past.

- **futMu**: an $M \times$ **horizonLong**-dimensional matrix, where M is the predictive sample size inherited from the **pandemic_model** function call and **horizonLong** is the input of the function. Each column is a sample of the predictive distribution of each mean function to the future.

- **fit**: the output of the **rstan::sampling** function.

- **seasonal_effect**: a vector of a minimum of zero (**NULL**) and a maximum of three strings, containing the requested weekdays for which a weekly seasonal effect is added to the daily counts mean in the model.

- **errors**: in case some samples of the posterior were removed due to them having absurd (cumulative counts larger than that location's population) values, then their indexes are returned in this element. Otherwise, an empty vector is returned. If such an exclusion happens, the function displays a warning message.

15.3.5 Calculating relevant statistics: pandemic_stats

Once a sample of the predicted values is obtained through the function **posterior_predict**, it is relatively straightforward to calculate point estimates and prediction intervals. The function **pandemic_stats** is responsible for returning the summary of all relevant statistics, such as short-term predictions, long-term predictions, estimated total number of cases, dates for the peak and end of the pandemic, and the median expected number of new observations $\mu(t)$.

Short-term predictions are calculated on the cumulative counts and long-term predictions are based on the new counts. The 95% credible interval is provided after calculating the mean, median, 2.5% and 97.5% percentiles for the predicted number of cases for each future date. The total number of cases is obtained by adding the cumulative total cases observed in the data to the predicted new cases for at least 1000 days ahead. The 95% credible interval takes into account the total number of cases considering the 2.5% and 97.5% percentile curves.

The peak of the pandemic curve represents the highest number of daily cases reported. As also mentioned in Chapter 10, this definition is important when considering a model with multiple waves. The median, 2.5% and 97.5% percentiles are calculated on the mean number of new cases $\mu(t)$ for the pandemic curve, starting from the first observed data point until at least 1000 days after the last date observed in the data. The 95% credible interval for the peak of cases is selected such that the two limiting dates of the 97.5% percentile curve coincide with the highest value of the 2.5% percentile curve. This guarantees that all possible curves belonging to the confidence band will peak within the defined interval.

Lastly, the dates indicating the end of the pandemic represent when the 99% percentile of the total number of cases was reached.

The resulting object of the function `pandemic_stats` belongs to the S3 class `pandemicStats`. It is a list containing the following elements:

- `data`: a list with a data frame containing the observed pandemic data, a string with the location name and a string indicating the type of cases predicted.

- `ST_predict`: a data frame with the short-term predictions for the number of cumulative cases. For each future date predicted, the mean, median, 2.5% and 97.5% percentiles are provided. The short-term horizon is determined by the `horizonShort` argument in the `posterior_predict` function.

- `LT_predict`: a data frame with the long-term predictions for the number of new cases. For each future date predicted, the mean, median, 2.5% and 97.5% percentiles are provided. The long-term horizon is determined by the `horizonLong` argument in the `posterior_predict` function.

- `LT_summary`: a list with the estimated total number of cases and the dates for the peak and end of the pandemic. In each metric, the median, 2.5% and 97.5% percentiles are provided.

- `mu`: a data frame with the median values of the mean number of new cases $\mu(t)$ for each date (starting from the first observed data point until the last date in the long-term horizon).

15.3.6 Plotting the results: `plot.pandemicPredicted`

The method `plot.pandemicPredicted` is responsible for the graphical visualisation of the predictions. Using the function `plot` with the object resulted from the `posterior_predict` function, as shown in Figure 15.1, the method generates interactive graphs containing the data, the predictions and the summary statistics of the pandemic for long and short terms. It is important to mention that the summary statistics are calculated internally on the method. However, if the user wants to save this information in an object, the function `pandemic_stats` can be used as explained in Section 15.3.5.

The documentation for the `plot.pandemicPredicted` method describes all possible arguments, which are as follows:

- `x`: an object of class `pandemicPredicted` created by the function `posterior_predict`.

- `term`: a string which indicates whether long-term, short-term, or both plots should be generated. The argument must be a string being either 'long', 'short' or 'both'. By default, only the long-term plot is generated.

- `color`: a logical variable indicating whether the plot should be coloured or in greyscale. If TRUE, then the plot is coloured, and grey otherwise. The default is a colour graph.

- summary: a logical variable indicating whether the plot should contain the summary statistics of the pandemic or not. If TRUE, then the plot shows the summaries, otherwise the plot does not show the notes. The default is TRUE.

It is possible to save the plots in an object. To do this, the user must assign the plot function with the desired settings to any variable. As a result, this variable will have two objects: short and long, containing, respectively, the short-term and long-term graphs. If any of them did not get plotted due to lack of prediction or due to the value of the term argument, its value will return NULL. The access to each plot is done in the regular way in R: variable$long and variable$short, where 'variable' is the name of the object that the plot was assigned to.

15.4 Modelling with the PandemicLP

In this section, the three different models currently contemplated by the PandemicLP package will be presented and illustrated. The selection of which model to fit occurs when assigning or failing to assign different inputs for the arguments n_waves and seasonal_effect in the pandemic_model function.

For the generalised logistic model, the default n_waves = 1 indicates that the model will be fitted to a curve with a single peak. When the user inputs n_waves = 2, the model chosen will be the two-wave model that allows for two peaks in modelling. Lastly, for the seasonal effect model, the user must choose at least one day of the week for the seasonal_effect argument. The model can be adjusted for both confirmed cases and deaths. The default case_type = "confirmed" indicates that the model will be fit to confirmed cases, whereas case_type = "deaths" uses the death cases for the model fit. More details on these choices will be discussed. For more in-depth theory behind each model and their properties, the reader should revisit Chapters 3, 4 and 6.

The following subsections illustrate, with COVID-19 data, the modelling of each of the three models contemplated in version 0.2.1 of the package. First, the user must load PandemicLP into R using the command library(PandemicLP). The listings included in this chapter contain the codes used for generating the outputs and figures presented here. To avoid confusion, the symbol > was added to differentiate a command line from an output, when necessary. To allow replication of the results by the user, an input seed is set in function pandemic_model. Further, for the purpose of this book, all graphs were created in greyscale using the command color = FALSE in the plot function.

15.4.1   Generalised logistic model

Commonly utilised when modelling the initial stages of a pandemic, the generalised logistic model (Equation (3.8)) is the simplest model available in the package. It is used to model regions that present only one peak. To explain, we will look at the number of deaths caused by COVID-19 in India from the beginning of the pandemic until the 12th of October 2020.

Listing 15.3 shows the code for how to pull COVID-19 data for India using load_covid and plot it with the function plot. For this example, our interest lies in modelling the number of deaths, thus, the number of confirmed cases from the plot will be omitted by clicking on 'Confirmed cases' from the legend. The resulting image is shown in Figure 15.2.

```
library(PandemicLP)
india <- load_covid(country_name = "india",
                    last_date = "2020-10-12")
plot(x = india, cases = "new", color = FALSE)
```

Listing 15.3: Loading and plotting COVID-19 data relative to India.

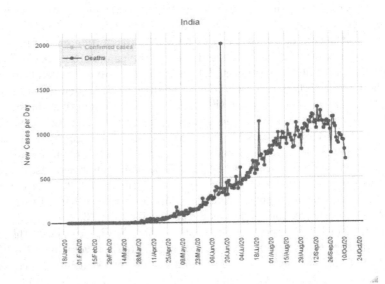

FIGURE 15.2: Number of new daily reported deaths in India from COVID-19 from the beginning of the pandemic until 12 [th] of October 2020.

As seen in Figure 15.2, the curve of daily reported deaths seems to have passed its peak and appears to be on a descending trend. Possible explanations for atypical observations such as the ones observed on 16th of June 2020 and 22nd of July 2020 have been explored in Section 12.2. No seasonal effect was identified, therefore the generalised logistic model should suffice. Listing 15.4 indicates how to fit the generalised logistic model with `pandemic_model` and how to print the returning object. The input argument `n_waves = 1` was specified in the code for emphasis despite being the default value.

```
 1 > fitindia <- pandemic_model(Y = india,
 2                              case_type = "deaths",
 3                              n_waves = 1,
 4                              seed = 123,
 5                              covidLPconfig = TRUE)
 6 > fitindia
 7 pandemic_model
 8  Distribution:          poisson
 9  Mean function form:    static generalized logistic
10  Type of Case:          deaths
11  Location:              India
12  Observations:          264
13
14 ------
15 Parameters:
16      mean    2.5%     50%    97.5%    n_eff    Rhat
17 a 44.305  19.027  42.016  80.146  549.468  1.000
18 b  0.495   0.427   0.497   0.548  494.448  1.001
19 c  0.013   0.013   0.013   0.014  490.424  0.999
20 f 12.223  10.991  12.258  13.265  504.571  1.002
21
22 ------
23 Priors:
24
25  a   ~   Gamma(0.1, 0.1)
26  b   ~   LogNormal(0, 20)
27  c   ~   Gamma(2, 9)
28  f   ~   Gamma(0.01, 0.01)
29
30 Restrictions:
31  1:  a/b^f < 0.02 *population
32  2:  f > 1
33
34 ------
35 *For help interpreting the printed output see ?print.
       pandemicEstimated
36 *For more information see ?'summary.pandemicEstimated
37 *For details on the model, priors and restrictions, see ?models
```

Listing 15.4: Fitting the generalised logistic model. The printed output gives information on the count distribution used, $\mu(t)$ form, parameter estimates, priors utilised, among other details specific to this model run.

Next, the `posterior_predict` function generates a predicted sample. The short-term forecast horizon was set to one week ahead, while the long-term forecast horizon was set to one year ahead. The function `pandemic_stats` contains all the main statistics of interest. Besides calculating point estimates and prediction intervals for the number of daily COVID-19 deaths in India in a year span, it informs the user about when the peak and end of the pandemic are estimated to occur. Alternatively, the user can also obtain this summary

from the graph as showed in the upper right corner of Figure 15.3. The R code
for each of those steps can be found in Listing 15.5.

```
1 > predindia <- posterior_predict(object = fitindia,
                                    horizonLong = 365,
2                                   horizonShort = 7)
3 > statsindia <- pandemic_stats(object = predindia)
4 > statsindia
5
6 95% Credible Intervals for death cases in India
7
8 Short-term Predictions:
9        date        q2.5         med        q97.5        mean
10 1 2020-10-13    110799.0    110856.0    110918.0    110857.2
11 2 2020-10-14    111766.9    111857.0    111948.0    111857.1
12 3 2020-10-15    112737.9    112854.0    112967.0    112851.5
13    ...
14 5 2020-10-17    114674.0    114827.0    114985.1    114828.6
15 6 2020-10-18    115634.0    115812.5    115983.0    115810.6
16 7 2020-10-19    116599.9    116793.0    116992.0    116792.7
17
18 ------
19 Long-term Predictions:
20        date      q2.5       med    q97.5       mean
21 1 2020-10-13   943.000   1000.0  1062.025  1001.201
22 2 2020-10-14   938.000    999.0  1063.000   999.857
23 3 2020-10-15   927.000    995.0  1057.000   994.459
24    ...
25 363 2021-10-10   9.000     17.0    26.000    16.977
26 364 2021-10-11  10.000     17.0    25.000    16.755
27 365 2021-10-12   9.000     16.0    25.000    16.556
28
29 ------
30 Total Number of Cases:
31      q2.5       med     q97.5
32   210142.9   222080   235189
33
34 ------
35 Peak Dates:
36      q2.5          med         q97.5
37   2020-09-06  2020-09-18  2020-10-03
38
39 ------
40 End Dates:
41      q2.5          med         q97.5
42   2021-08-23  2021-08-29  2021-09-04
43
44 ------
45 *Use plot() to see these statistics in a graph format.
46 *For more information, see ?'pandemicStats-xs'.
47 *For details on the calculations, see ?pandemic_stats.
48
49 > plotindia <- plot(x = predindia, term = "long", color = FALSE)
50 > plotindia$long
```

Listing 15.5: Making predictions, calculating relevant statistics and plotting
final results for the series of deaths in India caused by COVID-19.

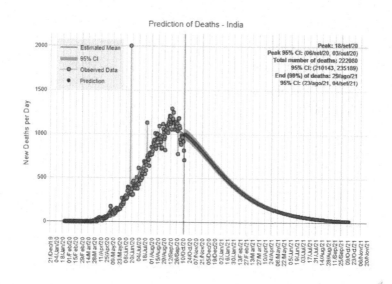

FIGURE 15.3: Long-term prediction plot for the series of deaths by COVID-19 in India made with data until 12th of October 2021. The vertical line indicates when the last observed data point happened. The series and points plotted can be identified with the help of the legend in the upper left corner. On the upper right corner, the 95% credible interval for the total number of deaths, and the peak and end dates for the pandemic are presented.

15.4.2 Generalised logistic model with seasonal effect

This subsection illustrates the use of the generalised logistic model for one wave with seasonal effect presented in Chapter 5 and described by Equation (5.2). The example data is the daily confirmed COVID-19 cases in the state of Minas Gerais, Brazil, from 25th of February 2020 until 15th of October 2020 used for model fitting, prediction calculations and visualisation of results (Listing 15.6 and Figure 15.4). It is important to note that version 0.2.1 of the package does not contemplate the inclusion of the seasonal effect for a model with more than one peak. To obtain the look of the plot in Figure 15.4, one must click in the 'Deaths' label in the legend to remove it from the original plot.

```
dataMG <- load_covid(country_name = "Brazil",
                     state_name = "MG", last_date = "2020-10-15")
plot(x = dataMG, color = FALSE) #new confirmed/deaths
```
Listing 15.6: Loading and plotting COVID-19 data from the state of Minas Gerais, Brazil.

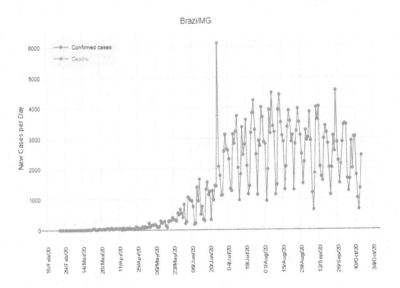

FIGURE 15.4: Time series plot of the confirmed daily cases of COVID-19 in the state of Minas Gerais, Brazil, from 25th of February 2020 until 15th of October 2020 (plot generated with code from Listing 15.6 after omitting the deaths time series in the `plotly` object).

Figure 15.4 demonstrates that fitting a generalised logistic model of one wave to the data seems reasonable. However, in this example, two days of the week will be considered to a have a weekly seasonal effect, specifically a pattern of delayed notification in both days: Sunday and Monday (for learning about the presence of seasonal effect, see Section 12.4). The model with seasonal effect is estimated by the function `pandemic_model` and requires the user to inform the input argument `seasonal_effect`. The argument `seasonal_effect` should receive the complete name, or the three first letters, of the two days of the week considered to have seasonal effect (input `seasonal_effect` supports a maximum of three weekdays). Listing 15.7 shows how the weekly seasonal effect is included in the `pandemic_model` call.

```
 1  > fitMG <- pandemic_model(Y = dataMG,
 2                            seasonal_effect = c("sunday", "monday"),
 3                            covidLPconfig = TRUE, seed = 123)
 4
 5  > fitMG
 6
 7  pandemic_model
 8    Distribution:           poisson
 9    Mean function form:     static seasonal generalized logistic
10    Type of Case:           confirmed
11    Location:               Brazil_MG
```

```
12  Seasonal effect:       sunday(d_1) monday(d_2)
13  Observations:          234
14
15  ------
16  Parameters:
17       mean  2.5%   50% 97.5%   n_eff   Rhat
18  a   8.083 7.198 8.087 9.070 801.998  1.000
19  b   0.117 0.107 0.117 0.128 658.915  1.003
20  c   0.022 0.021 0.022 0.022 683.481  1.004
21  f   5.127 4.875 5.127 5.385 658.003  1.003
22  d_1 0.795 0.787 0.795 0.804 857.173  1.001
23  d_2 0.431 0.426 0.431 0.437 908.121  1.000
24
25  ------
26  Priors:
27
28  a     ~    Gamma(0.1, 0.1)
29  b     ~    LogNormal(0, 20)
30  c     ~    Gamma(2, 9)
31  d_i   ~    Gamma(2,1)
32  f     ~    Gamma(0.01, 0.01)
33
34  Restrictions:
35  1:   a/b^f < 0.08 *population
36  2:   f > 1
37
38  ------
39  *For help interpreting the printed output see ?print.
         pandemicEstimated
40  *For more information see ?'summary.pandemicEstimated
41  *For details on the model, priors and restrictions, see ?models
```

Listing 15.7: Fitting the one-wave model with seasonal effect, applied on Sundays and Mondays, to the series of confirmed COVID-19 cases in the state of Minas Gerais, Brazil setting seed 123 for the generation of random numbers.

Observe in the print of object fitMG (on lines 22-23 of Listing 15.7) that the posterior point and interval estimates of parameters d_1 and d_2, concerning the Sunday and Monday effects respectively, are lower than 1, confirming the presence of the underreporting effect in those days. Furthermore, note that delayed notification effect on Monday $(42.6\% - 43.7\%)$ is greater than on Sunday $(78.7\% - 80.4\%)$. For more information about the fitting, use summary(fitMG).

The values for the diagnostic statistics n_eff, the effective sample size, and Rhat, the measure of comparison of between-and-within estimates (on lines 17-23 of Listing 15.7), are as expected for converge, i.e., high values for n_eff and Rhat less than 1.05. More information on these diagnostic measures is available in Chapter 6. In addition, the Stan sampler did not return any warnings about the lack of convergence or diagnostic problems.

Other convergence diagnostics can be used on the fit element of the output fitMG (see Listing 15.8 and Figure 15.5). In particular, the relevant parameters are a, b, c, f, d_1 and d_2. Note that the functions `traceplot` and `density` can be used directly with the `fitMG` output.

```
1 traceplot(fitMG) + theme(legend.position = "")
2 stan_ac(fitMG$fit, pars = c("a","b","c","f","d_1","d_2"))
3 density(fitMG)
```

Listing 15.8: Code for obtaining the convergence diagnostics plots shown in Figure 15.5.

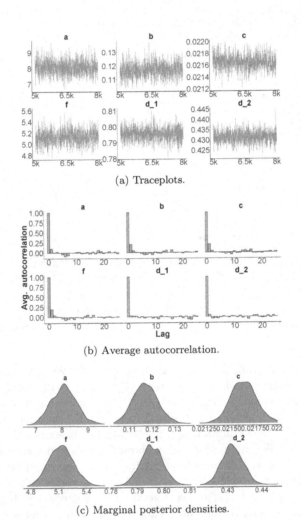

(a) Traceplots.

(b) Average autocorrelation.

(c) Marginal posterior densities.

FIGURE 15.5: Diagnostic analysis graphs for chains of the parameters for the model with seasonal effect applied to confirmed daily cases of COVID-19 in the state of Minas Gerais, Brazil, from 25th of February 2020 until 15th of October 2020. Subfigures generated from each command line from Listing 15.8.

Like before, for short and long-term prediction calculations, the generic function **posterior_predict** is used (Listing 15.9). The only compulsory input argument is the output of the function **pandemic_model** with the arguments **horizonLong = 500** and **horizonShort = 14** as default. Going back to the example of data from the Brazilian state of Minas Gerais, a long-term prediction window of 100 days ahead will be considered for better graphic visualisation of the pandemic behaviour without changing the short-term window.

```
1 > predMG <- posterior_predict(fitMG, horizonLong = 100)
2 > predMG
3
4 Predicted pandemic confirmed cases for Brazil_MG. Can be plotted
     with plot().
5
6 Showing predictive stats::median for the long-term predictions
     for Brazil_MG.
7
8 2020-10-16 2020-10-17 2020-10-18
9       2060       2031       1589
10 ...
11
12 2021-01-21 2021-01-22 2021-01-23
13     332.0      326.0      318.5
14
15 *For customized view, see help(print.pandemicPredicted)
16 **For more details, see help(pandemicPredicted-xs)
```

Listing 15.9: Making long-term and short-term predictions for the model with seasonal effect applied to the series of confirmed COVID-19 cases in the Brazilian state of Minas Gerais observed until 15th of October, 2020.

The output object **predMG** gives a small summary of the long-term predictions. The **print** method also allows visualisation of other statistics beyond the median default, as well as the summary of the short-term predictions. See package documentation **?print.pandemicPredicted**.

The function **pandemic_stats** provides a list with the point and interval posterior estimates for the short and long-term predictions, along with other useful statistics for understanding the behaviour of the pandemic: dates for the peak and end, and number of total cases at the end of the pandemic (see Section 15.3.5). In order to use the **pandemic_stats** function (Listing 15.10), the only input argument needed is the output of the generic function **posterior_predict**.

```
1 > statsMG <- pandemic_stats(predMG)
2 > statsMG
3
4 95% Credible Intervals for confirmed cases in Brazil_MG
5
6 Short-term Predictions:
7         date       q2.5        med      q97.5       mean
8 1 2020-10-16   330370.0   330462.0   330562.0   330461.8
9 2 2020-10-17   332368.0   332489.0   332621.0   332491.1
10 3 2020-10-18   333932.0   334079.5   334243.0   334080.2
```

```
11  ...
12  12 2020-10-27    347897.0    348236.0    348545.0    348227.9
13  13 2020-10-28    349569.9    349935.0    350274.0    349931.0
14  14 2020-10-29    351210.8    351607.0    351974.1    351605.9
15
16  ------
17  Long-term Predictions:
18          date      q2.5      med     q97.5      mean
19  1 2020-10-16 1968.000    2060.0 2160.050 2059.848
20  2 2020-10-17 1945.975    2031.0 2110.050 2029.266
21  3 2020-10-18 1514.950    1589.0 1668.025 1589.104
22  ...
23  98  2021-01-21  293.975    332.0  372.000  332.188
24  99  2021-01-22  289.975    326.0  365.025  325.246
25  100 2021-01-23  283.000    318.5  355.025  319.071
26
27  ------
28  Total Number of Cases:
29      q2.5     med      q97.5
30   420093.3 428471 437581.9
31
32  ------
33  Peak Dates:
34        q2.5          med        q97.5
35   2020-08-11 2020-08-18 2020-08-22
36
37  ------
38  End Dates:
39        q2.5          med        q97.5
40   2021-03-14 2021-03-17 2021-03-19
41
42  ------
43  *Use plot() to see these statistics in a graph format.
44  *For more information, see ?'pandemicStats-xs'.
45  *For details on the calculations, see ?pandemic_stats.
```

Listing 15.10: Calculating relevant statistics for the series of confirmed COVID-19 cases in the Brazilian state of Minas Gerais observed until 15th of October, 2020.

Lastly, the use of the `plot` method allows graphic visualisation of all results, short-term (Figure 15.7) and long-term predictions (Figure 15.6) inside an interactive `plotly` object. Figures 15.6 and 15.7 are static visual representations of the `plotly` object generated by Listing 15.11. The mandatory input argument is the output of the generic function `posterior_predict`.

```
1  plot(x = predMG, color = FALSE)  #default=plot long term
       predictions
2  plot(x = predMG, term = "short", color = FALSE) #plot short term
       predictions
```

Listing 15.11: Code for plotting final results for the model with seasonal effect applied to the series of confirmed COVID-19 cases in the Brazilian state of Minas Gerais shown in Figure 15.6 (long-term predictions) and Figure 15.7 (short-term predictions).

The visualisation of the seasonal effect is more evident when observing the estimated curve for the mean number of new cases, $\hat{\mu}(t)$ (Figure 15.6). It is possible to see that this curve shows a decay every Sunday or Monday, confirming the underreporting effect included in the modelling. An identical behaviour is also observed for long-term predictions (black dots in Figure 15.6). This feature can be easily seen by the user when manipulating the `plotly` interactive object generated by the command `plot(predMG)`.

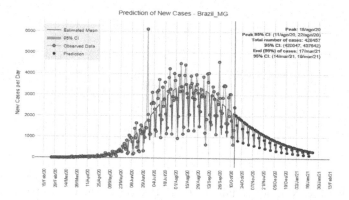

FIGURE 15.6: Estimated curve for the mean number of new cases, long-term predictions and estimates for dates of peak and end, total number of cases at the end pandemic, and respective 95% credibility intervals for the confirmed cases of COVID-19 in the state of Minas Gerais, Brazil. The last observed data point was in 15th of October, 2020 and the long-term prediction horizon was set to be 100 days ahead.

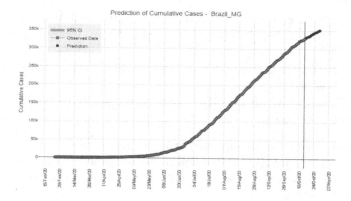

FIGURE 15.7: Short-term predictions and respective 95% credible intervals for confirmed COVID-19 cases in the state of Minas Gerais, Brazil. The last observed data point was in 15th of October, 2020 and the prediction horizon was set to be 14 days ahead.

15.4.3   Two-wave model

As discussed in Section 5.4, sometimes the pandemic has a second wave and this behaviour can be observed in the data. For example, we use the COVID-19 cases in the US with the data up to mid-July 2020, as shown in Figure 15.8.

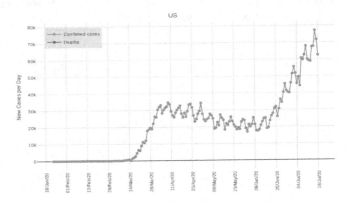

FIGURE 15.8: Number of daily COVID-19 cases in the US from the beginning of the pandemic until 18th of July 2020.

Fitting a two-wave model with the `PandemicLP` package is straightforward. The `n_waves` argument should be used for it. In particular, the model from Equations (5.6) and (5.7) is fitted. Unlike the generalised logistic model, there is no f parameter. Instead, each curve has two parameters: δ_j and α_j where $j = 1, 2$ represents each of the curves. The data can be fitted with the code in Listing 15.12.

```
1 us2Waves <- load_covid("US",last_date="2020-07-18")
2 us2W_est <- pandemic_model(us2Waves,n_waves=2,seed=123,
3                            covidLPconfig = TRUE)
4 density(us2W_est)
5 us2W_pred <- posterior_predict(us2W_est,horizonLong=100)
6 plot(us2W_pred)
```

Listing 15.12: Fitting US COVID-19 confirmed cases up to 18th of July 2020 with a two-wave model.

Figures 15.9 and 15.10 show the output of lines 6 and 8 of Listing 15.12. By using function `density`, the marginal posterior density of the model parameters are plotted. In this case, the model parameters are `aj`, `bj`, `cj`, `alphaj` and `deltaj`, where `j = 1, 2`.

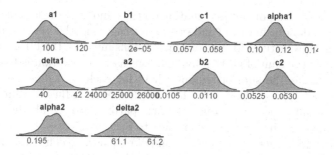

FIGURE 15.9: Density plot of the model parameters for the two waves model applied to COVID-19 US daily confirmed cases up to 18 of July 2020.

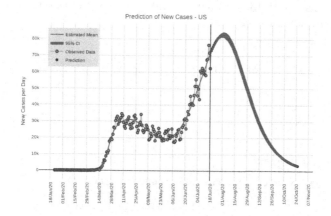

FIGURE 15.10: Prediction of the two waves model applied to COVID-19 US daily confirmed cases up to 18 of July 2020.

Plotting the result of the `posterior_predict` function is a simple way to see the prediction. The figure also provides credibility intervals for the prediction, although they are very small and practically invisible.

15.5 Sum of regions

As discussed in Section 5.2, sometimes, the need to model bigger areas as the sum of small ones may arise. For example, the user may need to predict the pandemic on a continent as the sum of the countries or model a country as the sum of the states. This is necessary when the evolution of the pandemic is heterogeneous through the areas. So, different regions can be in different stages of the pandemic, therefore, considering the data for only the bigger area might result in a loss of important information on the pandemic's true general status. This can be done using the `PandemicLP` package and this section presents a way to make predictions for two or more regions in which the future values are calculated as the sum of each individual forecast. This section is to exemplify the model fit as the sum of regions where for each we will run the default model of the package. However, the models can be adjusted differently for each individual region depending on their pattern. To see all the individual models available, the user can refer back to Section 15.4.

The first step when modelling the sum of regions is to define the vector that contains all the desired regions and the final date to be considered. Besides that, it is also necessary to specify if the modelling should be for the number of cases or deaths. To exemplify, this section will make the prediction for the South Region of Brazil COVID-19 deaths using the data until 1st of October, 2020. The Listing 15.13 contains the code to download the data and run the model separately for each state.

```
require(PandemicLP)

regions <- c("PR", "SC","RS") # list of states
last_date <- "2020-10-01"
case_type <- "deaths"

data <- list()
outputs <- list()
preds <- list()
states <- state_list()
for(i in 1:length(regions)) {
  if (is.na(match(regions[i],states$state_abb))){
    data[[i]] <- load_covid(country_name=regions[i],
                            last_date=last_date)
  } else {
    data[[i]] <- load_covid(country_name="Brazil",
                            state_name = regions[i],
                            last_date=last_date)
  }
  outputs[[i]] <- pandemic_model(data[[i]],case_type = case_type,
                                 covidLPconfig = TRUE, seed
    = 123)
  preds[[i]] <- posterior_predict(outputs[[i]])
}
```

Listing 15.13: First steps to run the sum of regions: load and modelling the multiple datasets

As presented in Listing 15.13, with the initial variables defined (lines 3 − 5), a loop is created to download the COVID-19 data from online repositories for each individual region using the function `load_covid`, and then run the models and make predictions using the functions `pandemic_model` and `posterior_predict`, respectively (lines 7 − 22).

Inside the loop, it is possible to see that an `ifelse` statement is done (lines 12 to 19) to check if the regions specified are Brazilian states. The check is done by comparing the regions provided with the available states, listed by function `state_list`. The data, outputs, and predictions for each region are stored in two lists (`outputs` and `preds`). The model is run using the argument `covidLPconfig = TRUE` at the `pandemic_model` function. Once again the user is encouraged to read Section 15.3 to know about all the arguments and functionalities of the available functions.

After running the model for each region, in order to make the predictions for the sum of them, some data manipulation is required and Listing 15.14 presents how to do it.

```
 1  data_base <- preds[[1]]
 2  bind_regions <- regions[1]
 3  for (i in 2:length(regions)) {
 4    bind_regions <- paste(bind_regions, "and", regions[i])
 5  }
 6  data_base$location <- bind_regions
 7
 8  # get the mean sample and set it to be dates x mcmc sample
 9  mu_t <- t(data_base$pastMu)
10
11  # include the dates in the data frame
12  mu_final <- data.frame(data = data_base$data$date,mu_t)
13  names_mu <- names(data_base$pastMu)
14
15  # get hidden objects (necessary for the pandemic_stats function)
16  hidden_short_total <- methods::slot(data_base$fit,"sim")$fullPred
          $thousandShortPred
17  hidden_long_total <- methods::slot(data_base$fit,"sim")$fullPred$
          thousandLongPred
18  hidden_mu_total <- methods::slot(data_base$fit,"sim")$fullPred$
          thousandMus
19
20  # loop for each region - starting with the second one
21  for (u in 2:length(regions)) {
22
23    # preds for the selected region
24    data_region <- preds[[u]]
25
26    # sum the variables predictive_Long, predictive_Short and futMu
27    data_base$predictive_Long <- data_base$predictive_Long +
28      data_region$predictive_Long
29    data_base$predictive_Short <- data_base$predictive_Short +
30      data_region$predictive_Short
31    data_base$futMu <- data_base$futMu + data_region$futMu
32
33    # create a large data frame by concatenating samples for
          current state in the mean data frame
34    mu_t <- t(data_region$pastMu)
35    mu_2 <- data.frame(data = data_region$data$date,mu_t)
36    names_mu <- c(names_mu,names(data_region$pastMu))
37    mu_final <- rbind(mu_final,mu_2)
38
39    # merge datasets by date since they can differ on start
40    data_base$data <- merge(data_base$data,data_region$data,
41                            by = "date", all = TRUE)
42    data_base$data[is.na(data_base$data)] = 0
```

```
43  data_base$data$cases.x = data_base$data$cases.x +
44    data_base$data$cases.y
45  data_base$data$deaths.x = data_base$data$deaths.x +
46    data_base$data$deaths.y
47  data_base$data$new_cases.x = data_base$data$new_cases.x +
48    data_base$data$new_cases.y
49  data_base$data$new_deaths.x = data_base$data$new_deaths.x +
50    data_base$data$new_deaths.y
51  data_base$data <- data_base$data[,-c(6:9)]
52  names(data_base$data) <- c("date","cases","deaths",
53                             "new_cases","new_deaths")
54
55  # sum hidden objects(necessary for the pandemic_stats function)
56  hidden_short_region <- methods::slot(data_region$fit,"sim")$
        fullPred$thousandShortPred
57  hidden_short_total <- hidden_short_total + hidden_short_region
58  hidden_long_region <- methods::slot(data_region$fit,"sim")$
        fullPred$thousandLongPred
59  hidden_long_total <- hidden_long_total + hidden_long_region
60  hidden_mu_region <- methods::slot(data_region$fit,"sim")$
        fullPred$thousandMus
61  hidden_mu_total <- hidden_mu_total + hidden_mu_region
62
63  }
64
65  # create hidden object(necessary for the pandemic_stats function)
66  methods::slot(data_base$fit,"sim")$fullPred$thousandShortPred <-
        hidden_short_total
67  methods::slot(data_base$fit,"sim")$fullPred$thousandLongPred <-
        hidden_long_total
68  methods::slot(data_base$fit,"sim")$fullPred$thousandMus <- hidden
        _mu_total
69
70  # aggregate the mean samples
71  mu_final <- aggregate(. ~ data, data=mu_final, FUN=sum)
72  mu_final <- mu_final[,-1]
73  mu_final <- t(mu_final)
74  names_mu <- unique(names_mu)
75  colnames(mu_final) <- names_mu
76  data_base$pastMu <- mu_final
```

Listing 15.14: Code to perform the sum of regions into a bigger one.

After fitting the individual regions, we want the final object to belong to the `pandemicPredicted` class and to contain the aggregated predictions, the data, the name of the regions merged, the type of data used, and the past and future μ's. So, to make sure that the object `data_base` reflects all the necessary format, we make it equal to the result of the `posterior_predict` function for the first listed region, and then the information will be added using a loop starting from the second index. This is done in lines 21 to 63 of Listing 15.14.

On the first step, the necessary objects are created by using the output of region 1 and creates the region name as the bind of each region (lines $1 - 6$). Then, in this case, the predictions (long and short) and the future μ's can be summed directly since the end date is the same for all regions, so the forecasts start at the same moment (lines 27 to 31). Now, for the sum of the data, however, it is necessary care because the dates of the beginning of the pandemic may be different for each region. So we must sum the cases only on common dates (lines 40 to 53). To do so, each data set is merged by date in a way that all the information is considered for both objects in the merge. When there is information on a given date for one region and not for another, the column with the missing value will be equal to NA, so it is necessary to change it to 0 to be correctly added. Then, each column from one data set is added to the corresponding column of the second one, respecting the dates difference. At the end, the duplicated column is deleted so that the loop can start again for another region without any problem.

In the case of the past μ, as shown in lines 9 to 13, the data frames containing the μ values are transposed in a way that each line represents a date and the columns are the values of each μ. After adding the column with the date, the data frame for each region is combined in only one (lines 34 to 37) and, after the loop, it is aggregated by date. The final data frame must have the date column deleted and needs to be transposed again to stay in the original format (lines 71 to 76).

Finally, to ensure that the **pandemic_stats** function has a horizon of data long enough to calculate pandemic summary information, some hidden objects are created by the **posterior_predict** function. These objects are generated considering the number of steps to predict to be equal to the maximum number between 1000 and the long-term horizon specified by the user. In order to do predictions as the sum of regions, it is important to sum these hidden objects as well (lines 16 to 18). Once again, they can be summed directly because all the data end on the same date, so the predictions are start at the same moment (lines 56 to 61).

Now, the object **data_base** contains all the necessary information to create plots, a summary, or any other analysis that the user may want. Figure 15.11 shows two prediction plots, one with the summary information (graph a) and a second one without the summary (graph b). These plots are obtained with the code in Listing 15.15, where lines 1 to 2 plot the prediction with the summary, and lines 3 to 5 without it by applying the argument **summary = FALSE** in the **plot** function.

```
1  plots <- plot(data_base, term = "long", color = FALSE)
2  plots$long
3  plots <- plot(data_base, term = "long", summary = FALSE,
4                color = FALSE)
5  plots$long
```

Listing 15.15: Plotting the merged results from different regions.

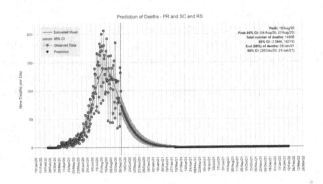

(a) Plot with the summary information.

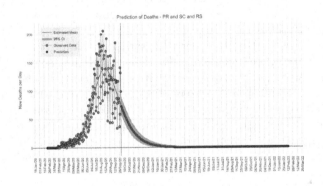

(b) Plot without the summary information.

FIGURE 15.11: Long-term prediction for the South region of Brazil done as the sum the states.

15.6   Working with user data

The purpose of this section is to show the user how to format the Y input argument in the `pandemic_model` function when epidemiological data was obtained outside of the `load_covid` function. COVID-19 data from cities, for example, are not supported by the package version 0.2.1.

The COVID-19 data for the city of Belo Horizonte, MG, Brazil will be used to illustrate how to correctly format the epidemiological data required by the `pandemic_model` function. The `covid19BH` data frame of the `PandemicLP` package contains the daily number of COVID-19 confirmed cases and deaths for the city of Belo Horizonte, from the date of the first notified case in 16th of March 2020 to 26th of June 2020. It has 103 observations and 6 variables in the following order:

1. `date`: dates in the YYYY-MM-DD format in descending order.

2. `new_confirmed`: number of new cases.

3. `new_deaths`: number of new deaths.

4. `last_available_confirmed`: cumulative number of cases.

5. `last_available_deaths`: cumulative number of deaths.

6. `estimated_population_2019`: size of Belo Horizonte's population.

The `covid19BH` data frame is from *Brasil IO* online repository (brasil.io/dataset/covid19, last accessed in 14th of November 2020) whose original sources are the State Health Secretariats in Brazil. Listing 15.16 is a partial view of this data frame.

```
> head(covid19BH)    # the first six lines
         date new_confirmed new_deaths
1 2020-06-26            35          3
2 2020-06-25           170          2
3 2020-06-24           105          8
4 2020-06-23           239          0
5 2020-06-22           345          0
6 2020-06-21            91          0
  last_available_confirmed last_available_deaths
1                     4977                   109
2                     4942                   106
3                     4772                   104
4                     4667                    96
5                     4428                    96
6                     4083                    96
  estimated_population_2019
1                   2512070
2                   2512070
3                   2512070
4                   2512070
5                   2512070
6                   2512070

> tail(covid19BH)       # the last six lines
           date new_confirmed new_deaths
98   2020-03-21            10          0
99   2020-03-20             2          0
100  2020-03-19             8          0
101  2020-03-18             5          0
102  2020-03-17             4          0
103  2020-03-16             1          0
```

```
32     last_available_confirmed last_available_deaths
33  98                        30                     0
34  99                        20                     0
35  100                       18                     0
36  101                       10                     0
37  102                        5                     0
38  103                        1                     0
39      estimated_population_2019
40  98                   2512070
41  99                   2512070
42  100                  2512070
43  101                  2512070
44  102                  2512070
45  103                  2512070
```

Listing 15.16: Partial view of the `covid19BH` data frame in `PandemicLP` package.

The data frame `covid19BH` has all the necessary information for fitting a growth curve model utilising function `pandemic_model`. However, this information is not in the required format. The data must be formatted in a list. The three elements of this list are as follows:

- `data`: a data frame with at least the following columns:

  - `date`: a date vector. It should be of class 'Date' and format 'YYYY-MM-DD'.
  - `cases`: a numeric vector with the time series values of the cumulative number of cases.
  - `new_cases`: a numeric vector with the time series values of the number of new confirmed cases.
  - `deaths`: a numeric vector with the time series values of the cumulative number of deaths.
  - `new_deaths`: a numeric vector with the time series values of the number of new deaths.

  This data frame should be ordered by date in ascending order.

- `name`: a string providing the name of Country/State/Location of the epidemiological data.

- *population*: a positive integer specifying the population size of the Country/State/Location selected.

Now, the goal is to build a list, named `dataBH`, in the format of the list required by the package `PandemicLP`, using the information in the `covid19BH` data frame. We initiate by building the element `data` in this list. For that to happen, it is necessary to rename the columns in the data frame as well as transforming the date vector into an object of the `Date` class. Since the covid19BH data frame is ordered by date in descending order (Listing 15.16), it is necessary to reorder the data frame in ascending order by date. Finally, it is possible to build a `dataBH` list by including the remaining elements `name` and `population`. Listing 15.17 presents all the necessary steps to transform the information in the covid19BH data frame into the dataBH list.

```
data <- covid19BH
names(data) <- c("date", "new_cases", "new_deaths", "cases",
                 "deaths", "population")
data$date <- as.Date(data$date)
data <- data[order(data$date, decreasing = FALSE), ]

pop <- data$population[1]
dataBH <- list(data = data, name = "Belo Horizonte/MG",
               population = pop) #the appropriate list
```
Listing 15.17: Building the input argument `Y` in `pandemic_model` using information of the covid19BH data frame.

The generic function `plot` allows visualisation of the data. However, first, it is necessary to set the list `dataBH` as an object of S3 class `pandemicData` (Listing 15.18). The resulting plot can be seen in Figure 15.12.

```
class(dataBH) <- "pandemicData"
plot(x = dataBH, color = FALSE) #new confirmed/deaths
```
Listing 15.18: Code to generate the graph with COVID-19 daily data from the city of Belo Horizonte, MG, Brazil. Source: covid19BH data frame from the **PandemicLP** package.

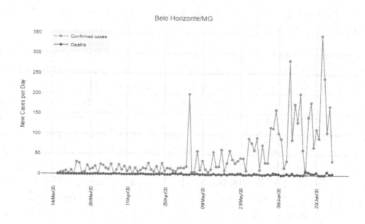

FIGURE 15.12: Time series plot of the number of daily COVID-19 cases, confirmed and deaths, in the city of Belo Horizonte, MG - Brazil, from the 16th of March until 26th of June 2020. Figure created with the commands in Listing 15.18.

Figure 15.12 demonstrates that the generalised logistic growth curve model with one wave can be a reasonable model for fitting the data for confirmed cases (the number of deaths observed is small and there is too little information in the data to provide reasonable fitting). If interested, the user can investigate the presence of a possible seasonal effect (see Section 12.4 for the motivation), and then adjust the seasonal effect model according to Section 15.4.2.

The next steps are fitting the one-wave model to the confirmed cases, making predictions and visualising the results. Listing 15.19 shows the command used and the output after fitting the one-wave model to the `dataBH` object, with the seed for generation of random numbers set to 123 for replication of results.

```
 1  > fitBH <- pandemic_model(Y = dataBH, seed = 123)
 2
 3  SAMPLING FOR MODEL 'poisson_static_generalized_logistic' NOW (
        CHAIN 1).
 4  Chain 1:
 5  ...
 6  Chain 1:
 7  Chain 1: Iteration:    1 / 5000 [  0%]  (Warmup)
 8  Chain 1: Iteration:  500 / 5000 [ 10%]  (Warmup)
 9  Chain 1: Iteration: 1000 / 5000 [ 20%]  (Warmup)
10  ...
11  Chain 1:
12  Warning messages:
13  1: There were 1 divergent transitions after warmup. Increasing
        adapt_delta above 0.8 may help. See
14  http://mc-stan.org/misc/warnings.html #divergent-transitions-
        after-warmup
15  2: Examine the pairs() plot to diagnose sampling problems
```

Listing 15.19: Generalised logistic model fit for confirmed cases of COVID-19 with data from list `dataBH`.

Note that the `Stan` sampler returned warnings about problems with the sampling (Listing 15.19, on lines 12-15), indicating a possible solution by using input arguments in the `Stan` sampler to control the sampling efficiency. For example, it suggests increasing the value of `adapt_delta`. The use of such settings in the `Stan` sampler will be explored in Chapter 16, although the setting `covidLPconfig=TRUE` of the function `pandemic_model` also contemplates some options for controlling the sampling efficiency. Generally, the use of `covidLPconfig=TRUE` tends to resolve problems with the posterior sample. The code in Listing 15.20 uses the setting `covidLPconfig=TRUE` with the purpose of controlling the sampling efficiency. The `Stan` sampler did not return any warnings about problems with the posterior sample and the diagnostic analysis from the model fit did not show any strong evidence of problems with convergence, sample independence or any other type of issue with the posterior sampling.

```
 1  > fitBHconfig <- pandemic_model(Y = dataBH, seed = 123,
 2                            covidLPconfig = TRUE)
 3
```

```
 4 SAMPLING FOR MODEL 'poisson_static_generalized_logistic' NOW (
       CHAIN 1).
 5 Chain 1:
 6 ...
 7 Chain 1:
 8 Chain 1: Iteration:    1 / 8000 [  0%]  (Warmup)
 9 Chain 1: Iteration:  800 / 8000 [ 10%]  (Warmup)
10 Chain 1: Iteration: 1600 / 8000 [ 20%]  (Warmup)
11 ...
12 Chain 1:
13
14 > fitBHconfig
15
16 pandemic_model
17  Distribution:          poisson
18  Mean function form:    static generalized logistic
19  Type of Case:          confirmed
20  Location:              Belo Horizonte/MG
21  Observations:          103
22
23 ------
24 Parameters:
25      mean     2.5%     50%    97.5%   n_eff    Rhat
26 a 145.287 126.069 145.218 165.971 992.034 1.002
27 b   0.003   0.001   0.002   0.010 422.355 0.999
28 c   0.031   0.025   0.032   0.035 398.347 0.999
29 f   1.140   1.005   1.104   1.470 387.564 0.999
30
31 ------
32 Priors:
33
34  a    ~   Gamma(0.1, 0.1)
35  b    ~   LogNormal(0, 20)
36  c    ~   Gamma(2, 9)
37  f    ~   Gamma(0.01, 0.01)
38
39 Restrictions:
40  1:  a/b^f < 0.08 *population
41  2:  f > 1
42
43 ------
44 *For help interpreting the printed output see ?print.
       pandemicEstimated
45 *For more information see ?'summary.pandemicEstimated
46 *For details on the model, priors and restrictions, see ?models
```

Listing 15.20: Generalised logistic model fit for confirmed cases of COVID-19 with data from list dataBH using setting covidLPconfig=TRUE.

Calculation of predictions, visualisation of results and graphs with MCMC sampling diagnostics are obtained using the commands from Listing 15.21. Figures 15.13 and 15.14 present the visualisation of results for the long- and short-term predictions respectively. Figures 15.15 present the diagnostics graphs from the MCMC method. From them we can see that it seems that the chain presented good convergence (Figures 15.15(a)) with small autocorrelation between the samples (Figures 15.15(b)). Finally, a marginal density plot (Figures 15.15(c)) is also visualised.

```
1  ## calculation of short e long-term predictions:
2  predBH <- posterior_predict(fitBHconfig)
3
4  ## calculation of relevant statistic for pandemic
5  statsBH <- pandemic_stats(predBH)
6
7  ## plot of the results:
8  plotBH <- plot(x = predBH, term = "both", color = FALSE)
9
10 plotBH$long     #plot long term predictions
11 plotBH$short    #plot short term predictions
12
13 ## convergence diagnostics graphics:
14 traceplot(fitBHconfig) + theme(legend.position = "")
15 density(fitBHconfig)
16 #autocorrelation plot:
17 stan_ac(fitBHconfig$fit, pars = c("a","b","c","f"))
```

Listing 15.21: Calculation of predictions, visualisation of results and convergence diagnostics for model fit from the object `fitBHconfig` obtained in Listing 15.20.

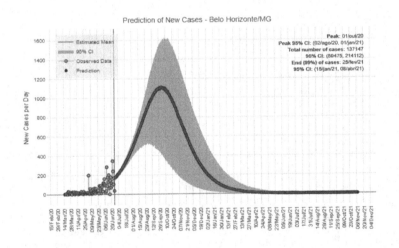

FIGURE 15.13: Estimated curve for the mean number of new cases, long-term predictions and estimates for the peak and end dates, as well as the number of total cases at the end of the pandemic and respective 95% credibility intervals for the confirmed COVID-19 cases in the city of Belo Horizonte, MG, Brazil. The last observed data point was on 26th of June 2020 and the prediction horizon was set to be 500 days ahead. Plot is contained within object `plotBH$long` from Listing 15.21.

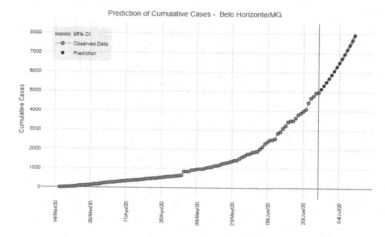

FIGURE 15.14: Short-term predictions and respective 95% credible intervals for the cumulative number of confirmed COVID-19 cases in the city of Belo Horizonte, MG, Brazil. The last observed data point was on 26th of June 2020 and the prediction horizon was set to be 14 days ahead. Plot is contained within object `plotBH$short` from Listing 15.21.

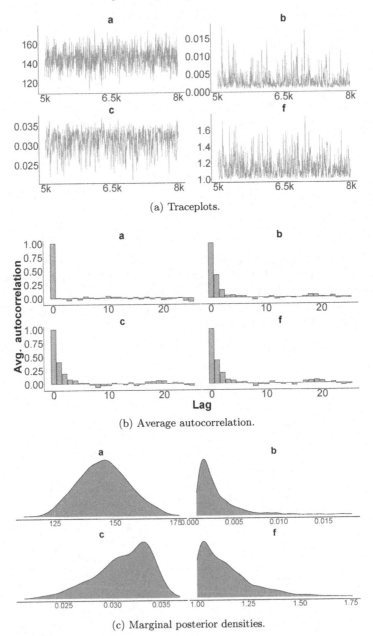

(a) Traceplots.

(b) Average autocorrelation.

(c) Marginal posterior densities.

FIGURE 15.15: Diagnostic analysis graphs for chains of the parameters for the generalised logistic one-wave model applied to confirmed COVID-19 cases in the city of Belo Horizonte, MG - Brazil, from 16th of March 2020 until 26th of June 2020. Plots generated by lines 14 - 16 in Listing 15.21.

Bibliography

Demarqui, F. N. (2020) *covid19br: Brazilian COVID-19 Pandemic Data.* URLhttps://CRAN.R-project.org/package=covid19br. R package version 0.0.1.

Gabry, J., Goodrich, B. and Lysy, M. (2020) *rstantools: Tools for Developing R Packages Interfacing with 'Stan'.* URLhttps://CRAN.R-project.org/package=rstantools. R package version 2.1.1.

Guidotti, E. and Ardia, D. (2020) COVID-19 data hub. *Journal of Open Source Software*, **5**, 2376.

Jenness, S. M., Goodreau, S. M. and Morris, M. (2018) EpiModel: An R package for mathematical modeling of infectious disease over networks. *Journal of Statistical Software*, **84**, 1–47.

Krispin, R. and Byrnes, J. (2020) *coronavirus: The 2019 Novel Coronavirus COVID-19 (2019-nCoV) Dataset.* URLhttps://CRAN.R-project.org/package=coronavirus. R package version 0.3.0.

Ooms, J. (2020) *V8: Embedded JavaScript and WebAssembly Engine for R.* URLhttps://CRAN.R-project.org/package=V8. R package version 3.4.0.

Ponce, M. and Sandhel, A. (2020) covid19.analytics: An R package to obtain, analyze and visualize data from the Corona Virus disease pandemic. *preprint.* URLhttps://arxiv.org/abs/2009.01091.

R Core Team (2020) *R: A Language and Environment for Statistical Computing.* R Foundation for Statistical Computing, Vienna, Austria. URLhttps://www.R-project.org/.

Stan Development Team (2020) RStan: The R interface to Stan. URLhttp://mc-stan.org/. R package version 2.21.2.

Wickham, H., Hester, J. and Chang, W. (2020) *devtools: Tools to Make Developing R Packages Easier.* URLhttps://CRAN.R-project.org/package=devtools. R package version 2.3.2.

16

Advanced settings: The `pandemic_model` function

Marcos O. Prates

Universidade Federal de Minas Gerais, Brazil

Guido A. Moreira

Universidade Federal de Minas Gerais, Brazil

Marta Cristina C. Bianchi

Universidade Federal de Minas Gerais, Brazil

Débora F. Magalhães

Universidade Federal de Minas Gerais, Brazil

Thais P. Menezes

University College Dublin, Ireland

CONTENTS

Chapter 15 presented how the `pandemic_model` function can be used to fit all the models available on version 0.2.1 of the `PandemicLP` package. The illustrated adjustments considered the settings of the MCMC sampler as the values used by the CovidLP app, whose specifications are accessed by the argument `covidLPconfig = TRUE`. Although the use of such an argument makes it easier for the user to run one of the available models, it may not guarantee that the generation of the model parameter chains will achieve convergence (and other desired properties in a sample via MCMC), especially in the case of fitting epidemiological data sets other than COVID-19. Thus, the main goal of this chapter is to illustrate another way to use the function `pandemic_model` by presenting some of its advanced settings for purposes of controlling the

DOI: 10.1201/9781003148883-16

efficiency of sampling via MCMC. In the sequence, considering a real example, this chapter presents a workflow for analyzing diagnostic sampling via MCMC using the tools available in the package. The last section discusses the use of the p input of the `pandemic_model` function, inherent in the truncation of the prior distribution of one of the model parameters for the purpose of alleviating problems with the predictions.

16.1   Introduction

For the implementation of sampling using the Markov Chain Monte Carlo method (MCMC) (or Hamiltonian Monte Carlo (HMC) and other variants that the user can check in Section 6.2.2), it is required to specify some basic parameters such as the number of iterations, warm-up, and sample length, parameters that are also responsible for the sampling efficiency. In the `PandemicLP` package, the function `pandemic_model` has the argument `covidLPconfig=TRUE` which is a predefined set of values for some of these elements used by the CovidLP project app. The `covidLPconfig=TRUE` default configuration is reasonable for providing generation of chains efficiently reaching convergence for COVID-19 data, the main subject of the CovidLP project. However, for other types of epidemiological data there is no guarantee that this configuration will provide success in posterior sampling. In this way, the user may need to understand the basic elements to run an MCMC chain (or HMC) and know how to manipulate them when using the function `pandemic_model`. To help in this, the main goal of this chapter is to provide the user, through examples using real data from COVID-19, the ability to manipulate these input arguments that are responsible for controlling the efficiency of sampling via MCMC. For that, it is necessary that the user already has knowledge about generic issues related to the implementation of an MCMC method, and thus once again, the user is invited to read Section 6.2.2. Besides that, this chapter will also provide ways to deal with possible sampling issues and will show how to do the diagnostic analysis of the MCMC chain using the tools in the package.

The sampling efficiency can be controlled when using at least one of the following input arguments of the `pandemic_model` function:

- `chains`: a positive integer specifying the number of Markov chains to run. The default is 1.

- `warmup`: a positive integer specifying the number of warm-up (also known as burn-in) iterations per chain. These warm-up samples are not used for inference. The default is 2000. If using the argument `covidLPconfig=TRUE`, then the default is changed to 5000.

- `thin`: a positive integer specifying the period for saving samples. The default is 3.

- `sample_size`: a positive integer specifying the posterior sample's size per chain that will be used for inference. The default is 1000. The total number of iterations per chain is `warmup + thin * sample_size`.

- `init`: specification of the initial values of the parameters per chain. The default is `random`. When using the argument `covidLPconfig=TRUE`, initial values for each parameter of the chosen model are pre-specified by the CovidLP team; these values are suitable for the COVID-19 data. Any parameters whose values are not specified will receive initial values generated as described in `init=random`. Specification of the initial values for `pandemic_model` can only be done via a list equal in length to the number of chains. The elements of this list should themselves be named lists, where each of these named lists has the name of a parameter and is used to specify the initial values for that parameter for the corresponding chain.

- `control`: a named list of parameters to control the `Stan` sampler's behavior. It defaults to NULL so all the default values of the sampler are used. Two of the possible adaptation parameters are used in this chapter:

  - `adapt_delta` (double, between 0 and 1, defaults is 0.8)
  - `max_treedepth` (integer, positive, default is 10)

  If using the argument `covidLPconfig=TRUE`, the default value of this parameter is `control = list(max_treedepth = 50, adat_delta = 0.999)`.

- `p`: a numerical value greater than 0 and less than or equal to 1. It is the percentage of the maximum cumulative total number of cases until the end of the epidemic in relation to the population of the location. The default is `p = 0.08`. However, using the arguments `covidLPconfig=TRUE` and `case_type=deaths`, the default becomes $p = 0.02$. This quantity refers to the prior distribution truncation inherent in the modeling proposed in this book. The interpretation and the theoretical reasoning about this percentage are discussed in Sections 6.1.2 and 13.2.

Of all the arguments presented above, only `p` does not directly influence the sampling efficiency of the MCMC method. As a matter of fact, the value of `p` can influence predictions calculated by the model on the long term, being able to solve problems related to the generation of excessively large values for the forecasts. The use of this input is covered in Section 16.4.

On the examples presented in this chapter, the one-wave generalised logistic model without seasonal effect will be considered for the sake of simplification. However, all the configurations illustrated here are applicable to any other model adjusted by the function `pandemic_model`. All the examples show how to achieve some desirable (or mandatory) characteristics in sampling via MCMC, like chain convergence and obtaining independent samples (approximately), prioritising, when possible, computational efficiency. Several of the examples in this chapter are based on the cases presented during Chapter 15 in order to discuss other aspects (regarding sampling efficiency) of these respective adjusted models.

Since the main goal is to discuss the aspects associated with the chains generated by the function `pandemic_model`, if it is of the user's interest to save such chains in an object in R, this task can be done using the functions `as.data.frame` or `as.matrix` on the `fit` element of the `pandemic_model` function output.

To ensure that the examples in this chapter can be reproduced by the user, it is important to highlight that the `seed` input argument was fixed for the generation of random numbers. The graphic methods of diagnosis of the MCMC sampling included in the package have the coloured default, however following the proposal of this book, their respective figures are presented here in the black and white version. To run the codes presented on this chapter, the `PandemicLP` package must be installed and loaded as explained in Section 15.2. The version of the package used in this chapter is 0.2.1.

16.2 Solving sampling issues

The main objective of this section is to use the function `pandemic_model` to generate chains for the different parameters of the model without notifications of any sampling issues by the `Stan` sampler and with reasonable computational efficiency. Reaching the given objective is analogous to reaching sampling efficiency via MCMC. To this end, one or more input arguments presented in Section 16.1 are used. They are responsible for the settings of the `Stan` sampler in the function `pandemic_model`. It is worth mentioning that values specified by the user for such input arguments are replaced by values specified by the standard settings of the CovidLP app if the option `covidLPconfig=TRUE` is active.

One of the main sampling issues of MCMC algorithms is often the lack of chain convergence. The simplest strategy is to increase the number of iterations in the chains. When necessary, this strategy is pointed out in the notifications of the `Stan` sampler. In that case, it is recommended that the user increase the initial chain period or warm-up period (samples that are not used for inference), changing the input argument `warmup` whose default value is 2000 iterations.

Since increasing the number of iterations will likely increase the computational time for completing the sampling, another strategy to improve convergence is to specify adequate initial values for the chains of every parameter in the model instead of them being generated randomly (`init="random"` default). Such strategy can help reduce the computational time due to the chains being initiated at values with higher probability of the posterior distribution, therefore, more rapidly covering the parameter space. To illustrate such strategy, we will return to the generalised logistic model fitting for the number of COVID-19 deaths observed in India until 12th of October 2020 (Listing 15.4 from Section 15.4.1). The model fit object will be updated to include data until the 31st of October, and we will use the estimated values (medians) for the parameters obtained previously as the initial values for the chains of every parameter in the model (code in Listing 16.1). The initial values for the chains are specified in the `init` input argument as a list. The set of initial values for the chains of each parameter of interest in the model, in this case a, b, c and f, are specified in a list whose elements are named as the respective parameter names (for specifying initial values for more than one chain see Section 16.3).

```
1  #loading data:
2  indiaData <- load_covid(country_name = "india",
3                          last_date = "2020-10-31")
4  #fitting:
5  fitindia_random <- pandemic_model(Y = indiaData,
6                                    case_type = "deaths",
7                                    warmup = 6000, seed = 123)
8  #including initial values for chains:
9  fitindia_init <- pandemic_model(Y = indiaData,
10     case_type = "deaths",
11     warmup = 6000, seed = 123,
12     init = list(list(a = 42.02, b = 0.497, c = 0.013, f = 12.25)))
```

Listing 16.1: Loading COVID-19 data from India and fitting model setting initial values for chains.

Regarding the sampler settings, the two model adjustments in Listing 16.1 differ only because the second model fit (`fitindia_init`) utilises good initial values for chains of the parameters in the model, whereas the first model fit (`fitindia_random`) uses random starting values. Both fittings generated chains with 9000 iterations each: 6000 iterations in the warm-up and after that, a posterior sample size of 1000 (default value for `sample_size` input) is created and saved with a period of 3 (default value for `thin` input). There were no warnings related to sampling issues. In this example, the inclusion of good initial values for the chains reduced almost 40% the sampling computational time (time savings depend on the machine's configuration).

In addition to the chain convergence problem, other important issues in sampling may exist. In this section, interest lies in resolving problems notified by the `Stan` sampler (after sampling is completed) that may interfere directly in the validity and efficiency of the inference made. More specifically, our main interest is to discuss two types of `Stan` notifications: "divergent transitions after warmup" and "transitions after warmup that exceeded the maximum treedepth". The first type of notification informs that the sampling does not guarantee the validity of the inference, whereas the second type concerns a possible loss of efficiency in the inference. When there is notification of at least one of these problems after sampling is completed, it is recommended that the input argument `control` is used to specify greater values than the default ones for the following adaptation parameters in Stan: `adapt_delta` (between 0 and 1, default is 0.8) and `max_treedepth` (positive integer, default is 10). As an example, going back to Section 15.6 we can see in Listing 15.19 that Stan returned warnings about the sampling of the generalised logistic model for confirmed COVID-19 cases in the city of Belo Horizonte, MG – Brazil. One of the notification messages is transcribed in Listing 16.2 below. This notification indicates that increasing the parameter value in `adapt_delta` might solve the issue. Thereby, an example of settings adopted in the function `pandemic_model` aiming to solve this problem is presented in Listing 16.3. In Section 15.6, the model fit presented in Listing 15.20 also solves this sampling issue by adopting the standard settings `covidLPconfig=TRUE`, because this option contemplates the use of the argument `control` to increase the value of the parameter `adapt_delta`.

```
1  Warning messages:
2  1: There were 1 divergent transitions after warmup.
3  Increasing adapt_delta above 0.8 may help.
4  See http://mc-stan.org/misc/warnings.html
5  #divergent-transitions-after-warmup
```

Listing 16.2: Warning messages of the Stan sampler about sampling problems in the model fit of confirmed cases from list dataBH from Section 15.6.

```
1  fitBH_control <- pandemic_model(Y = dataBH, seed=123,
2                             control = list(adapt_delta = 0.95))
```

Listing 16.3: Using advanced settings to control the sampler's efficiency (with the input argument `control`) in the model fit to confirmed cases from list dataBH from Section 15.6.

The sampling issues mentioned in the previous paragraph refer to the behaviour of the sampling method used in the sampler of function `pandemic_model`. It is a variant of the HMC method (for more about Hamiltonian Monte Carlo, see Section 6.2.2) called the No-U-Turn-Sampler or "NUTS". As detailed in Section 6.2.2, the HMC method (and its variants) contribute to improve the performance of the MCMC sampler through its potential of generating chains that can cover all the parameter space in the posterior distribution in a more efficient way than the other MCMC methods. More theoretical details about this method go beyond the scope of this book. The Stan Manual (`mc-stan.org/users/documentation/`, last accessed on 13th of December 2020) presents, in an intuitive manner, the behaviour of the sampler when using the "NUTS" method with values higher than the default for the adaptation parameters `adapt_delta` and `max_treedepth`:

... imagine walking down a steep mountain. If you take too big of a step you will fall, but if you can take very tiny steps you might be able to make your way down the mountain, albeit very slowly. Similarly, we can tell Stan to take smaller steps around the posterior distribution, which (in some but not all cases) can help avoid these divergences.

In this way, requiring more control over the behaviour of the sampler, the sampling becomes more conservative and robust, although it becomes slower. However, in the presence of divergent transitions, the priority is the validity and efficiency of the inference rather than computational time.

Even after the solution of possible chain convergence problems and other sampling issues related to the validity and efficiency of the inference, the MCMC method can still have generated chains with high autocorrelation. Chains with this characteristic produce samples potentially concentrated in specific parts of the parameter space, and thus are not a good representation of the posterior distribution. However, based on the observation of the chain autocorrelation graphs, it is possible to verify if the simple strategy of thinning the posterior samples (samples taken after warm-up, used for inference) can lessen the problem of high autocorrelation.

```
COLdata <- load_covid(country_name = "colombia",
                      last_date = "2020-07-15")

fitCOL <- pandemic_model(Y = COLdata, case_type = "deaths",
            warmup = 6000, seed = 123,
            control = list(max_treedepth = 40, adapt_delta = 0.95))

fitCOL_10 <- pandemic_model(Y = COLdata, case_type = "deaths",
            warmup = 6000, thin = 10, seed = 123,
            control = list(max_treedepth = 40, adapt_delta = 0.95))
```

Listing 16.4: Loading and fitting models for COVID-19 deaths from Colombia from the first case observed until 15th of July 2020.

Highly autocorrelated chains and a possible solution for guaranteeing (approximate) sample independence are illustrated in Listing 16.4. The model `fitCOL` exemplifies the generation of highly autocorrelated chains for the parameters of the generalised logistic model for COVID-19 deaths in Colombia since the first case observed until 15th of July 2020. The model `fitCOL_10` presents a (probable) solution for guaranteeing (approximate) sample independence by saving samples with a greater lag.

Figure 16.1 was created by Listing 16.5 and shows sampling diagnostic graphs for each of the models. Neither traceplots (Panel 16.1a and Panel 16.1c) presents strong evidence for lack of convergence. Yet the traceplots, as well as the autocorrelation graphs, (Panel 16.1a and Panel 16.1b, respectively) show that the chains generated by model `fitCOL` are highly correlated. The traceplots present strong evidence of high autocorrelation because consecutive iterations have generated values close to each other. In addition, in Panel 16.1b note that the default value `thin=3`, presents an average autocorrelation around 0.5 for parameters b, c, and f, and that the autocorrelation starts to decrease as of lag 10. The Panel 16.1c and Panel 16.1d present the diagnostic graphs for chains generated by a model with increased thinning (`thin=10`). In particular, the chain autocorrelation graph (Panel 16.1d) shows low values, indicating that lag 10 is reasonable for saving posterior samples and guaranteeing (approximate) sample independence. Lastly, it is worth mentioning that the models `fitCOL` and `fitCOL_10` did not present any warnings in the `Stan` sampler about sampling issues indicating the need to use the configuration argument `control` to provide more efficient sampling.

```
1 #convergence diagnostics for fitCOL
2 traceplot(fitCOL) + theme(legend.position = " ")
3 stan_ac(fitCOL$fit, pars = c("a", "b", "c", "f"))
4 #convergence diagnostics for fitCOL_10
5 traceplot(fitCOL_10)
6 stan_ac(fitCOL_10$fit, pars = c("a", "b", "c", "f"))
```

Listing 16.5: Convergence diagnostics for the generalised logistic model applied to COVID-19 deaths from Colombia from the first case until 15th of July 2020.

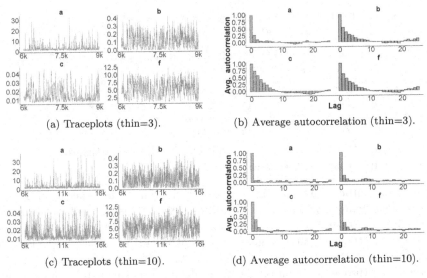

(a) Traceplots (thin=3).   (b) Average autocorrelation (thin=3).

(c) Traceplots (thin=10).   (d) Average autocorrelation (thin=10).

FIGURE 16.1: Diagnostic analysis graphs for chains of the parameters for the generalised logistic model applied to the series of COVID-19 deaths in Colombia from the first case until 15th of July 2020. Panel 16.1a and Panel 16.1b present the traceplot and the average autocorrelation for the model fit using the default `thin=3`, respectively. Panel 16.1c and Panel 16.1d present the traceplot and the average autocorrelation for the model fit using `thin=10`, respectively.

16.3   Sampling diagnostics

One of the main things to note when performing Bayesian inference using MCMC, presented in Section 6.2.2, is that it is based on a theoretical guarantee that the sampled values are distributed according to the posterior distribution as the number of iterations approach infinity. In a real scenario, drawing an infinitely large sample is unfeasible. Instead, a finite sized sample is drawn and the sample is considered to be drawn from the posterior if enough evidence that the chain has converged to its stationary distribution can be found.

Package `rstan` provides a few tools which help detect when a Markov Chain has converged. This section will present practical examples of the use of such tools. One of them is fitting multiple chains with different and independent starting values. The idea behind this is that the chains should converge to sampling from the posterior regardless of the starting values. If multiple chains converge to the same distribution, then this increases the chance that it is the posterior. A safe and recommended choice is to use 4 independent chains.

The first and least formal convergence diagnostic tool is the traceplot. It is a line plot of the sampled values against the chain iterations. If the sampled values vary around the same region in the parameter space, then there is evidence of convergence. If multiple chains do so in the same region, then it is further, stronger evidence.

Another important tool is the $\hat{R}$ statistic, which benefits from multiple chains. It makes a comparison of the average "within-chain" variance with the "between-chain" variance. If they are consistent with one another, then it is likely that the chains have converged as well. It is presented by (Gelman and Rubin, 1992) and indicates convergence if it is less than 1.01.

A further important tool is the effective sample size, denoted n_{eff}. In some cases, the Markov Chain is autocorrelated, meaning there is correlation of the values of the chain with lagged versions of itself. Autocorrelation after convergence means that some information is "repeated" among sampled values, which leads to estimation with less information than would come from an independent sample of the same size. Then n_{eff} represents the size of an independent sample which would yield an equivalent result of the correlated n-sized sample.

Finally, a useful statistic is the Monte Carlo standard error. It is the estimated error for estimations based on the MCMC sample. It can be used to detect when more draws of the posterior are necessary for estimation. Note that this is not a convergence diagnostic, but it is mostly analysed during this phase as it may represent the need for more iterations from the algorithm.

To illustrate the sample diagnostic analysis done by the function `pandemic_model`, using the tools available in `PandemicLP` package, the example of the generalised logistic model for COVID-19 deaths in India will be used again considering the data until the 31st of October of 2020. This data is accessed as shown in Listing 16.1 of Section 16.2. In this section, however, instead of sampling only 1 chain (`chains = 1` is the default), Listing 16.6 samples 4 chains for each model parameter, with different initial values specified for each chain (respecting the prior knowledge of the parametric space of the parameters of interest). The `init` input of `pandemic_model` function is an object of the class "list". The specification of different initial values for different chains (or just accepting the default `init="random"`) is important for the diagnostic analysis of sampling, as each chain starting at a different point (or location) of the parametric space adds greater reliability to the convergence range.

```
1 #specifying initial values for the four chains:
2 init4 <- list()
3 for(i in 1:4){
4 init4[[i]] <- list(a = 42.02+(i-1), b = 0.497+(i-1)/10,
5                    c = 0.013+(i-1)/100 , f = 12.25-(i-1))
6 }
7 #fitting model
8 fitindia_chains <- pandemic_model(Y = indiaData,
9                     case_type = "deaths", warmup = 6000,
10                    chains = 4, init = init4, seed = 123)
```

Listing 16.6: Fitting model to the COVID-19 deaths regarding data from list `indiaData`, loaded by Listing 16.1, using four chains with different specified initial values.

As shown in Listing 16.7, the `summary` function presents basic information about the data, the chosen model and sampler settings, as well as posterior estimates for some quantities of interest regarding the model parameters (in this case, a, b, c, e f and new cases mean $\mu(t)$, $t = 1, \ldots, n$, where n is the size of the observed data). In this section, however, the interest is to explore only the sampling diagnostic statistics presented in the package in a brief and practical way for the example `fitindia_chains`. Listing 16.6 shows the code for this, to avoid confusion, the symbol $>$ was added to differentiate a command line from an output.

```
1  >summary(fitindia_chains)
2  >pandemic_model
3   Distribution:          poisson
4   Mean function form:    static generalized logistic
5   Type of Case:          deaths
6   Location:              India
7   Observations:          283
8
9   ------
10  Parameters:
11     mean se_mean      sd   2.5%    50% 97.5%     n_eff   Rhat
12  a 6.598   0.021   0.888  5.026  6.531 8.448 1774.322  1.002
13  b 0.037   0.000   0.004  0.029  0.037 0.046 1561.865  1.001
14  c 0.020   0.000   0.000  0.019  0.020 0.020 1556.464  1.001
15  f 3.068   0.004   0.149  2.791  3.062 3.366 1585.768  1.001
16
```

```
17 Fitted values:
18        mean    sd   2.5%   50%  97.5%
19 mu[1]   0.365 0.044 0.285 0.362 0.455
20 mu[2]   0.386 0.046 0.303 0.383 0.481
21 mu[3]   0.409 0.048 0.322 0.406 0.507
22         ...   ...   ...   ...   ...
23 mu[281] 695.308 6.025 683.695 695.280 707.269
24 mu[282] 686.888 6.123 675.108 686.891 699.082
25 mu[283] 678.479 6.220 666.485 678.461 690.869
26
27 ------
28 covidLPconfig =  FALSE :
29
30  warmup:                               6000
31  thin:                                 3
32  sample_size:                          1000
33  chains:                               4
34  maximum total number of cases:        0.08 *population
35  init (chain_id = 1):
36 $a
37 [1] 42.02
38
39 $c
40 [1] 0.013
41
42 $f
43 [1] 12.25
44
45 $b
46 [1] 0.497
47
48 ------
49 Priors:
50
51  a  ~  Gamma(0.1, 0.1)
52  b  ~  LogNormal(0, 20)
53  c  ~  Gamma(2, 9)
54  f  ~  Gamma(0.01, 0.01)
55
56 Restrictions:
57  1:  a/b^f < 0.08*population
58  2:  f > 1
59
60 ------
61 *For help interpreting the printed output see ?summary.
         pandemicEstimated
62 *For more information see ?'summary,stanfit-method'
63 *For details on the model, priors and restrictions, see ?models
```

Listing 16.7: Summary for model fitindia_chains fitted by Listing 16.6.

Listing 16.7 presents the fit summary for model fitindia_chains, where the diagnostic statistics don't present strong evidence for any problems with the model parameters. The Rhat statistics are less than 1.01 as recommended by the Stan Team, showing that the four chains mix well. The Monte Carlo standard error (se_mean) for each parameter is small, guaranteeing that estimations have small approximation error. An interesting result is that the effective sample size is actually larger than the sample size. This can happen when the Markov chain has very little autocorrelation and the marginal posterior is approximately Gaussian.

Although diagnostic statistics quantify and summarise information about sampling validity and efficiency in an effective manner, graphical methods are often richer in providing information for diagnostic analysis of MCMC sampling. Figure 16.2 created in Listing 16.8 presents the three plotting methods included in the package to assist in the sampling diagnosis of the model fit to the COVID-19 data from India. As with the diagnostic statistics, the plot methods also do not present evidence of any sampling problems. Panel 16.2a shows traceplots of the four chains for each model parameter. It is visually clear that for each parameter, the chains mix well being located under the same region of the parametric space, thus showing the reach of convergence. Panel 16.2b presents the mean autocorrelation plot of the samples, leading to the conclusion that considering `thin = 3` (default) is sufficient to guarantee approximate sampling independence. The posterior marginal density presented in Panel 16.2c is unimodal and well behaved.

```
1  traceplot(fitindia_chains)
2  density(fitindia_chains)
3  stan_ac(fitindia_chains$fit, pars = c("a", "b", "c", "f"))
```

Listing 16.8: Convergence diagnostics for the generalised logistic one-wave model applied to COVID-19 deaths from India from the first case until the 31st of October 2020.

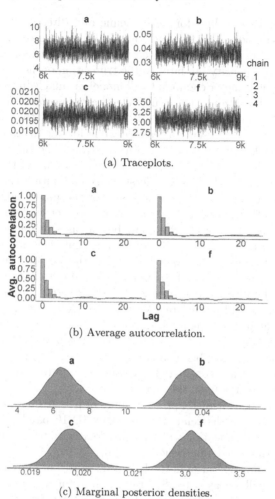

(a) Traceplots.

(b) Average autocorrelation.

(c) Marginal posterior densities.

FIGURE 16.2: Diagnostic analysis graphs for chains of the parameters for the generalised logistic one-wave model applied to COVID-19 deaths from India from the first case until the 31st of October 2020.

16.4   Truncation of the total number of cases

The input p of the `pandemic_model` function presented in Section 16.1 is not related to any settings of the MCMC sampler. However, the models proposed by the CovidLP Team (including the models available in version 0.2.1 of the package) can overestimate the predicted values (point and interval estimates) producing unrealistic values for an epidemic and directly affecting the efficiency of the sampler. This fact has already been discussed in Section 6.1.2 and Section 13.2 and can be associated with the problems of model identifiability. Hence, the quantity p relates to the truncation of the prior distribution of parameter a (parameter common to all models available in the package) and has the role of diminishing the occurrence of extreme values in the sampling procedure controlling for a more realistic prediction of the model. In other words, this quantity represents a feasible value for the maximum fraction of cases in the population of the region of interest until the end of the epidemic, represented as $p \times N$, where $0 < p \leq 1$ and N is the size of the region's population. Consequently, the reasonable strategy of limiting the total number of cases predicted by the model to the size of the population of the region of interest (or a fraction of it) controls, in an indirect way, the generation of values excessively large for the daily number of cases predicted.

The problem of generating excessively large numbers for the predictions (point and interval estimates) was observed in the CovidLP app particularly when the models were adjusted to epidemiological data in the early stages of the pandemic (data from the beginning of the time series). However, the reason for that problem is still not entirely known. Therefore, regarding the choice of a value for input p in the `pandemic_model` function, it is recommended that the user first attempt to use the default value (p = 0.08) if this is considered feasible for the epidemic being analysed. In case the predictions obtained by the fitted model are considered excessively large (or even unrealistic), it is recommended to readjust the model with a lower value of p. It is worth mentioning that the CovidLP app uses values of p = 0.08 for confirmed cases and p = 0.02 for deaths related to the COVID-19 pandemic. There is no guarantee that this truncation will solve problems with the predictions because the reason for this modelling issue is not fully known.

Data from confirmed cases in the US until the 20th of March 2020 presented in Figure 16.3 will be used to illustrate a situation where predictions with excessively large values, considered to be unrealistic, were obtained. This data represents the beginning of the time series for COVID-19 confirmed cases, since the first notified case in the US occurred on the 23rd of January 2020 (only two months' worth of observations).

Two settings of the generalised logistic model fit to US data, using p = 1 (without restriction) and the default value p = 0.08, are discussed. It is important to remember that the argument p refers to the restriction of the number of cases predicted by the fitted model at the end of the pandemic being, at most, $p.100\%$ of the size of the US population ($326, 687, 501$ people in 2018, information extracted by function `load_covid`). The code for loading data, fitting the model, calculating predictions and visualising the results is presented in Listing 16.9. There were no warnings in `Stan` about sampling issues in either one of the models fitted.

```
1  #loading data:
2  USdata <- load_covid(country_name = "us",
```

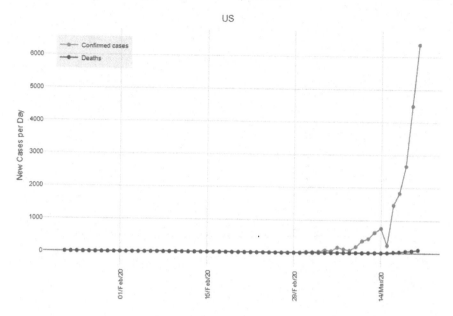

FIGURE 16.3: Number of daily COVID-19 cases in the US from the beginning of the pandemic until 20th of March 2020.

```
3                   last_date = "2020-03-20")
4  #fitting model
5  fitUS_p1 <- pandemic_model(Y = USdata, p = 1,
6                     warmup = 6000, seed = 123,
7                     control = list(max_treedepth = 50,
8                     adapt_delta = 0.99))
9
10 fitUS_p08 <- pandemic_model(Y = USdata,
11                     warmup = 6000, seed = 123,
12                     control = list(max_treedepth = 50,
13                     adapt_delta = 0.99))
14 #posterior sample, long and short-term predictions and plot:
15 predUS_p1 <- posterior_predict(fitUS_p1)
16 plot(predUS_p1)
17 predUS_p08 <- posterior_predict(fitUS_p08)
18 plot(predUS_p08)
```

Listing 16.9: Loading and fitting models to data of COVID-19 confirmed cases in the US from the first case until 20th of March 2020.

Figure 16.4 shows each of the models adjusted with their respective predictions. Panel 16.4a pertains to the model fit using `p` = 1 whereas Panel 16.4b refers to the use of `p` = 0.08. Note that both models produce predictions for the daily number of cases that are absurd, reaching point estimates around 10 million daily cases when using `p` = 1, and 1.5 million for `p` = 0.08. The upper limit of the credibility interval for the total number of cases at the end

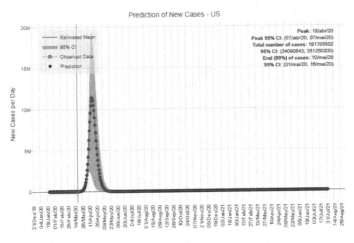

(a) Model fit with **p** = 1.

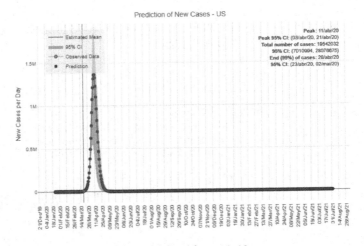

(b) Model fit with **p** = 0.08.

FIGURE 16.4: Estimated curve for the mean number of new cases, long-term predictions and estimates for the peak and end dates as well as the number of total cases at the end of the pandemic and respective 95% credibility intervals for the confirmed COVID-19 cases in the US. The last observed data point was in 20th of March 2020 and the prediction horizon was set to be 500 days ahead. Panel 16.4a presents the results of the model fit with p=1, whereas Panel 16.4b presents the results for the default input p=0.08.

of the pandemic predicted by the model fit with p = 1 exceeds the population size specified in the model. However, when using the default value p = 0.08, the upper limit for the total number of cases at the end of the pandemic is predicted to be around the 8% specified in the model fit, thus deemed admissible. Although both settings have generated predictions with excessively large values, by using a lower value (feasible) for the specification of p, the results obtained were more reasonable.

Bibliography

Gelman, A. and Rubin, D. B. (1992) Inference from iterative simulation using multiple sequences. *Statistical Science*, **7**, 457–472.

Part VII

Conclusion

17

Future directions

The CovidLP Team

Universidade Federal de Minas Gerais, Brazil

CONTENTS

Summary

This chapter outlines final thoughts on possible ways to proceed from what the book described. Ideas on how to proceed are presented for all main areas of the project: modelling, monitoring, implementation and software.

17.1 Introduction

A large number of issues were considered by our team during the last year, during the development of the CovidLP Project. Many of them were successfully operationalised and implemented, but not all of them. A few of them are singled out in this final part of the book as our primary items in a virtual to-do list. They are succinctly presented in the same order of the parts of the book: modelling, monitoring, implementation and software.

DOI: 10.1201/9781003148883-17

17.2   Modelling

Many other modelling ideas can still be implemented beyond the ones presented in this book. Among them, a few were selected for further thoughts here, mostly because they were all entertained by the CovidLP team during the course of the project. These will be presented in the sequel.

17.2.1   Overdispersion

As discussed in Chapter 4, pandemic counts of daily new cases and deaths are usually overdispersed with respect to the Poisson distribution. As was also discussed, there are countless ways to add probabilistic overdispersion to a model. The few ways discussed in this book are achieved by assigning a distribution to the Poisson rate parameter, so that the marginal expected value for the data is the desired mean curve. The special case where this distribution is Gamma implies a marginal Negative Binomial distribution for the data. Two of its parametrisations were explored and were referred to as the mean-dependent and the mean-independent overdispersion. The case where the Poisson rate parameter is distributed according to log-Normal and log-Skew-Normal distributions were also explored.

Although these options are robust, none of them showed reasonably good results with the COVID-19 data when the counts had not reached their respective peaks. Predictions often underestimated future counts with the point-wise prediction and the credible intervals were too large. A possible way to treat this was also discussed in Section 4.3, which is to include a prior truncation of the overdispersion parameter so that an intermediate point between the overdispersed and the Poisson models can be achieved. However, there is no guarantee that there is a guideline for creating a reasonable truncation value. A clear future direction in this sense is to explore parametric or non-parametric options to add overdispersion in the model.

17.2.2   Relation between cases and deaths

An interesting point to consider is the relation between confirmed cases and confirmed deaths. It is reasonable to assume that the total number of confirmed deaths that occurred at a time t_d, was once solely confirmed cases in a time $t_c \in (0, t_d)$–for each confirmed death there should be a date of confirmation of the disease case. Therefore, one may make an effort to model this relation and bring it into the model. Motivations to follow this strategy would be, for instance

- to consider a limiting value for the number of deaths at any time t, as it cannot be greater than the total confirmed cases *until* the same time point;

- it can help estimation and prediction of the confirmed deaths over time, since one would have the information of the confirmed cases of which a percentage can eventually become death cases;

- to evaluate the mortality rate over time, which may decrease/increase due to reasons like the overload of health care systems and deeper understanding of the disease.

However, despite these ideas, there are a few difficulties in applying this strategy. Here, we cite the increase in model complexity and that the available data series may be indexed only with the reporting date, instead of the date of first symptom or any disease-related date. As extensively discussed in this book, the reporting date may be influenced by bureaucratic and other factors that are not related to the disease under study, e.g., delay in reporting. Therefore, the relation between cases and deaths may be ultimately too complicated to be modelled. Furthermore, it may be necessary to have a well-structured data base at subject level.

17.2.3   Automated identification of wave changes

Some techniques to model pandemic counts with multiple waves were proposed in Section 5.4. These techniques require a previous decision about the number of curves to be considered in the model. As will be mentioned in Section 17.4, a decision about how many waves exist in daily counts of cases or deaths in pandemic data, based only on visual inspection, is a subjective choice. It may be useful to replace this visual inspection strategy by an automated decision procedure. This procedure may be designed, for example, under some methodology of change point identification. A wide range of methods for change point identification can be found in Chen and Gupta (2012), Brodsky and Darkhovsky (1993), Brodsky (2016) and others. The change point model should be able to point out that the typical decreasing or stable behaviour observed after the counts peak has probably changed to a new stage of increasing trend of these counts. When a new change is identified, the model may be extended to consider a more suitable structure that will provide better predictions of the future behaviour of the pandemic counts, which are now in a new increasing stage. For example, for the model described by Equations (5.6) and (5.7), an automated decision procedure would provide an appropriate number of curves to be considered, without the need of any subjective strategy.

17.3   Implementation

It is the job of the implementation team to execute the steps required to automate and publish any new extensions of the project. Even without changes in the methodology, there are tasks and improvements that must be kept on our radar as we have an online platform to maintain. These tasks include making sure that the most up-to-date and complete data sources are always accessed, updating the list of regions being analysed, and monitoring the automated processes performed every day to fit all the prediction models.

Regarding the data sources used, challenges with interruption of access and lack of updates from official predictions were already faced. Even a small delay in the data publishing can cause prediction to become outdated, since automated scripts for model fit are scheduled to be automatically launched daily. Thus, any prediction platform needs to be alert to these possible issues to be able to readily overcome them when needed.

A probable modification in the near future is to change the way we access the data for Brazilian states. Right now, an intermediary repository that extracts the data from the Ministry of Health is used, as mentioned in Chapter 9. We are working on having alternative tools to download the data directly, and this may open the possibility of having data on other levels (such as cities) for Brazil.

Since our team is mostly based in Brazil, we have turned our efforts in maintaining the most complete modelling approach for our country. However, the relevance of applying other model extensions to more countries, such as weekly seasonal effect and sum of states for large countries is acknowledged. We will continue to study and monitor the pandemic situation, and make these changes as we are capable.

As the model complexity increases, so does the run time needed for proper estimation. This time can often be longer than the 24-hour window that is initially set for the scripts to start running for daily update, causing queuing problems as delays in the most up-to-date information at the app. For this reason, to keep improving the aspects of platform models, it might be necessary to review and change the periodicity at which results are updated.

Due to the time and workforce constraints of the CovidLP team, some visual features have not yet been implemented in our app at the time this book is being written. So far, we have prioritised enabling timely predictions for a vast list of countries and Brazilian states, as well as making the presentation of the long-term forecasts as clean as possible for the wide range of backgrounds of the public. However, there are some possible interesting additions to the app on our list that we would like to detail here.

In the first tab of the app, where we show the graphs of the observed series of new and cumulative cases/deaths, it could help the user to understand the current pandemic situation to include more options for viewing the data. For example, the option to select the series of number of cases and deaths relative to the region's population, as well as selecting more than one region to compare their data.

On the other tabs, where we show the model fit and forecast results, some possible extensions include calculating and displaying the reproduction number R. This quantity indicates how contagious the disease currently is, and it can be mathematically defined based on the logistic model we presented in the previous chapters.

Since our results are updated frequently, it would be interesting to visualise the change in predictions for a given region over time, as more data points are being observed and taken into the model. For this purpose, it would be necessary to go back in time and fit the same current model to every single day since the beginning of the pandemic and save all the sample results. Currently, we only store the last daily results to avoid a memory burden.

17.4 Monitoring

This section suggests some ideas for future directions to improve the work of monitoring the behaviour of epidemic data through time. Unfortunately, it was not possible to implement these ideas in the current version of the book, but the discussion presented here should be considered in the development of new tools intended to guide decisions from the monitoring team in other epidemiological applications. Recall that the periodic (possibly daily) monitoring of an epidemic has a central role in the type of study proposed in this book. The identification of atypical observations and possible behaviour in the serial data, such as seasonality and waves, are necessary to choose the most appropriate statistical model.

The seasonality modelling in this book was applied to the Brazilian data sets, but other countries may also exhibit a similar pattern of high/low counts in regular cycles of the 7 weekdays for the COVID-19 data. This pattern may not appear in the entire time series for a given region. For example, the weekend effect does not occur at the beginning of the series, but it can be detected after a few months. The monitoring team is expected to daily update the list of regions displaying this seasonality, which means that the adopted non-seasonal statistical model should be replaced for some of the series. In the CovidLP project, seasonality was basically identified through visual inspection of the count series. Particularly, the seasonality was evaluated using boxplots for each day of the week, but this is still an empirical analysis requiring visual judgement from the analyst. As a future step to simplify this evaluation, a strategy should be developed, perhaps expressing uncertainty through probabilities, to support or refute the hypothesis that a weekend effect is present in the serial data. Naturally, this strategy should consider data from at least a few seasonal cycles in order to indicate high probability of weekend effect.

Decisions based on a visual inspection can be subjective and they require discussions between the members of the monitoring team to avoid adopting the biased view of a single person. Developing a tool to make this type of decision less dependent on subjective opinions is definitely an important task.

In the CovidLP project, the conclusions related to the number of waves and about the moment in which an additional wave begins were also based on visual inspection. Different countries or states show distinct configurations of these features in their time series. Creating an automated strategy expressing probabilities to support or refute the hypothesis about the presence of an extra wave is a proposal to be considered in future research. Again, the monitoring team is expected to daily update the list of regions displaying multiple waves, and then indicate to the implementation team the series for which the statistical model should change to handle the new data pattern. However, the reader must have in mind that fitting a model assuming multiple waves has a higher computational cost. Therefore, the daily update of predictions for a large number of regions is a challenge in this case. Any method designed to identify additional waves in an epidemic time series should account for the fact that a delay in data addition may occur, due to technical problems, leading to the incorporation of the corresponding values a few days later, together with the new observations. This means an inflated count for the days after the delay period, which does not represent the beginning of an extra wave. The detection of an extra wave must consider an increasing trend related to several days.

17.5 Software

Software development implies continuously improving the existing code. As such, it is very likely that by the time this book is available, improvements have already been made to the code for the PandemicLP package in R discussed in Chapters 15 and 16. Nonetheless, the pursued improvements are listed here. The most important long-term goal of the CovidLP team for the package is to provide continuous support, bug corrections and occasional updates. As highlighted in Chapter 15, the package can be used to make predictions for any pandemic or epidemic daily counts data. Not only can the package be used for past pandemics, but the package could be useful for long-term predictions in possible future epidemics as well. For this reason, the team aims to keep the package working and updated.

The highest priority in short-term improvement is the implementation of various models that were discussed throughout this book. Some of those are already being used in the shiny app discussed in Chapter 14, available online in https://dest-ufmg.shinyapps.io/app_COVID19/. The models which are expected to be included in the short term are as follows:

- k-curves mean for $k \geq 2$ (multi-wave mean) with the Poisson distribution for the data

- multi-curves mean with seasonal effects with the Poisson distribution

- negative binomial distribution for the data (both parametrisations discussed in Chapter 4)

- negative binomial models with $k \geq 2$ multi-curves mean

- negative binomial models with seasonal effects

- negative binomial models with $k \geq 2$ multi-curves mean and seasonal effects

New models may appear and they might be implemented in the future as well. In addition to the new models, functionality improvements are expected in the short term. As it is, the `pandemic_model` function requires the data to be inputted in a very specific format. Although this format is explained in detail in the package documentation, a function will be added to allow the user to loosely input the data and output it in an adequate format for the `pandemic_model` function. Another current limitation of the `pandemic_model` function is that it requires both confirmed cases and deaths to be informed, even if the user only wishes to model one of them. This restriction will be changed to allow for data that have only one of these responses.

The package currently has two vignettes, which can be viewed in `https://cran.r-project.org/web/packages/PandemicLP/vignettes/PandemicLP.html` and `https://cran.r-project.org/web/packages/PandemicLP/vignettes/PandemicLP_SumRegions.html`. They respectively exemplify the basic usage of the package and a way for the user to make . predictions of the sum of regions with it. Another vignette will be added showing a short tutorial of how to format custom data so that it can be used in the `pandemic_model` function.

The latter vignette will become much simpler once the formatting function is implemented. Additionally, the code for the sum of regions plots the data withholding the caption which shows some summary statistics in the plot. This happens due to the `pandemic_stats` calculating the summary statistics by using hidden data in the prediction objects and this treatment is not corrected in the vignette. Therefore, a function will be added which joins predictions for different regions into a single object with the adequate predictions for their sum.

Finally, a further improvement in the package roadmap is to allow advanced users to change the prior distribution of the parameters, not just their class of distributions, in the `pandemic_model` function. This will allow the user to add information to the model as discussed in Section 6.1.2. The default values currently implemented were found to be adequate for the COVID-19 data, but may not be adequate for data from another pandemic.

Bibliography

Brodsky, B. (2016) *Change-Point Analysis in Nonstationary Stochastic Models*. New York: Chapman and Hall/CRC, 1st edn.

Brodsky, E. and Darkhovsky, B. S. (1993) *Nonparametric Methods in Change Point Problems*, vol. 243 of *Mathematics and Its Applications*. Springer Netherlands, 1st edn.

Chen, J. and Gupta, A. K. (2012) *Parametric Statistical Change Point Analysis: With Applications to Genetics, Medicine, and Finance*. Birkhäuser Basel, 2 edn.

Index

Printed in the United States
by Baker & Taylor Publisher Services